Principles of Paint Formulation

Edited by

R. WOODBRIDGE
Formerly Technical Director of Berger Decorative Paints
Bristol
and
Past President of the Paint Research Association
Teddington, Middlesex

Blackie
Glasgow and London

Published in the USA by
Chapman and Hall
New York

Blackie & Son Limited
Bishopbriggs, Glasgow G64 2NZ
and
7 Leicester Place, London WC2H 7BP

Published in the USA by
Chapman and Hall
a division of Routledge, Chapman and Hall, Inc.
29 West 35th Street, New York, NY 10001–2291

First published 1991

British Library Cataloguing in Publication Data

Principles of paint formulation.
I. Woodbridge, R.
667.01

ISBN 0-216-93144-4

Library of Congress Cataloging-in-Publication Data

Principles of paint formulation / edited by R. Woodbridge.
p. cm.
Includes bibliographical references and index.
ISBN 0-412-02951-0
1. Paint. I. Woodbridge, R. (Richard)
TP935.P73 1991
667'.6—dc20 91-16426
CIP

Phototypesetting by Pure Tech Corporation, Pondicherry, India
Printed in Great Britain by Thomson Litho Ltd, East Kilbride, Scotland

Preface

During the last century the art of paintmaking has progressed at an ever increasing pace from a craft to a science. But within one aspect of paint technology, the art of formulation, some craft attitudes still prevail. Review of the wealth of technical literature available today identifies a paucity of information available to the newcomer to the industry, or to the practising paint technologist who wishes to improve his skills, on how to approach the challenge of reformulation or the development of a new product. Quite naturally, a paint manufacturer will not disclose his formulations to a competitor, but the basic principles of formulation should be common to all. Guidance on this has been restricted to the recommendations of raw material suppliers who, not surprisingly, bias their investigations and recommendations towards highlighting the benefits of their products. This is not a criticism of their activities – indeed, without their contributions, formulation of paint might still largely be in the hands of craftsmen, albeit scientifically trained craftsmen who had developed or inherited their skills.

This volume aims to set out the basic principles of formulating paint in a competitive commercial environment, which should be applicable to most types of liquid surface coatings. The procedures recommended in the early chapters have been developed and used over many years and have withstood the test of time. Their aim is to save work and time. They have been used within the development laboratories of a major paint manufacturer in a flexible way, being adapted according to the specific needs of individual projects. Other companies may require a different approach, but the same principles are likely to apply. Only by careful consideration of objectives and by adequate planning at all stages will the desired result be achieved in the most cost-effective way.

A book on paint technology would not be complete without some basic information on the raw materials available and their correct selection, which is the subject of Chapter 3. Subsequent chapters by co-authors, recognised as specialists in their fields, concentrate on specific aspects of paint formulation, such as Fred Stieg's review of the importance of volume relationships and ways in which simple techniques can assist in approaching reformulation in a more scientific way. For anyone involved in, or about to consider, computer techniques, Luigi Cutrone's chapter on the use of computers is essential reading, covering the basic design of experiments through to the latest 4-component contour plotting technique in formulation development. Ike Berg adds his years of experience in adapting formulations from the laboratory scale to bulk manufacture, with guidelines on correct operating techniques for most types of equipment. The book also includes a chapter on the use of alternative 'dry hiding' technologies as a means of reducing usage of titanium dioxide, and concludes with guidance on the selection of test methods and some useful appendices on paint calculations, international units and conversions, health and safety, and a glossary of terms.

To all authors I extend a special thanks, not only for their willingness to contribute, but for their forbearance in responding to suggestions for minor changes or additions. I also wish to thank members of the staff of the Paint Research Association at Teddington – particularly John Bernie, Len O'Neil, Dip Dasgupta, Norman Morgan, Len Tysall and Catherine Hayworth, for their helpful advice, retrieval of information and references, general encouragement and support, and for permission to republish the 'Units for the Paint Formulator' and the 'Glossary of Terms'.

R.W.

Contributors

Ken Arbuckle Consultant, 20 Oak Lodge Avenue, Chigwell, Essex IG7 5HZ, UK

Ike Berg Consultant, 22 Gladstone Court, Anson Road, London NW2 4LA, UK

Luigi Cutrone Tioxide Italia SPA, Stabilimento di Scarlino, Contrada Casone CP113, 58022 Fallonica, Italy

Fred Stieg Pigmentech, 903 Beachview Drive, Jekyll Island, California, USA

Richard Woodbridge Consultant, Rest Harrow, Staunton Lane, Whitchurch, Bristol BS14 0QG, UK

Contents

1 Getting the basics right

RICHARD WOODBRIDGE

1.1 Introduction

This chapter concentrates on the development of methods of working which establish a basic structured background to any development project. The aim is to encourage a disciplined approach which should lead to more efficient work practices and hence to more effective formulation.

Clear and unambiguous laboratory records must be maintained in order to allow easy retrieval of information. It is strongly recommended that a loose-leaf system is adopted rather than the alternative diary approach. This system enables the incorporation of preprinted standard forms, for the entry of formulations and results, into the finished project file (see Section 1.5).

1.2 Availability of raw material data

Raw materials are the building blocks of any paint formulation. Selection of the correct grades is fundamental to the development of a balanced product within the identified cost target. Guidance on the correct selection and optimum levels of usage will be given in Chapter 3. Selection implies that the formulator has at his or her fingertips, if not already in mind, all necessary information on the available raw materials appropriate to the particular formulation being developed. The relevant information is specific to individual countries and indeed to individual companies within each country. The availability of some resins or latices at realistic cost levels may be influenced by the commercial availability of intermediates or monomers to the resin producer. These differences in availability can influence significantly the relative cost effectiveness of different grades of raw materials, leading to positive differences in the 'local' approach to formulation. As examples of this, the early growth of latex paints in the United States was based on styrene butadiene, while in the United Kingdom early developments were based on vinyl acetate homopolymers. Even in the development of vinyl copolymers differences in the comonomer used in world markets, or differences in the relative costs of vinyl copolymers and 100% acrylics are found. Similarly, in the field of extenders wide differences in the grades available and their costs, depending on indigenous supplies, are also found. Such factors make the range of 'available raw materials' specific to each country.

What information, therefore, is required by the formulator and should be immediately accessible to ensure correct choice? It will vary according to the class of raw materials, but should include the following basic data.

Chemical composition is required to establish any possible interactions or influence on final product performance. It is also essential to enable the technologist to assess any potential hazard in the storage or handling of the raw material or the finished product. From this aspect it is necessary to be aware of the presence of any minor constituents which may influence the overall degree of risk in handling or use. While such information was not always readily available, current health and safety regulations and increasing concern for the environment place a statutory obligation on suppliers and users to disclose, or to be aware of, the nature of any risks based on the evidence of composition.

Physical properties are also important indicators of which grade of many raw materials are suitable for a particular application. In the case of pigments or extenders, information on properties such as mean particle size and size range (not necessarily full particle size distribution), oil absorption and specific gravity are required. For coloured pigments an indication of light fastness and durability, etc., is important. Data on resin solutions or latices should include solid content, specific gravity of both the solution and the solid resin and viscosity data. For comparative purposes it is useful to have an indication of solids content at a standard low viscosity – 200 cP is often quoted. In the case of resins supplied in low boiling point solvents, flash point is a critical property. For solid resins, melting point may be a useful record. For solvents specific gravity, boiling point or range, evaporation rate, and flash point should be recorded.

Most of this information is readily available from the supplier's literature, together with a mass of other information on every product in their range, all presented in individual styles ranging from simple printed sheets to voluminous glossy brochures. In this form it is not readily available to the formulator and, more importantly, in technical data sheets there is no reference to cost. Cost information is important to the formulator if targets are to be met but is subject to negotiation on quantity purchased and varies from one company to another. For most 'dry' raw materials, cost is quoted on a weight basis but since most paint products are sold on a volume basis, the formulator's target is to achieve a given cost per litre or gallon. Particularly in the case of extenders it is important to recognise the cost contribution of a unit *volume* of extender towards the total product cost. This may give a very different viewpoint on the relative costs of extenders being considered. This consideration has a significant impact on the approach to formulation of a product traditionally sold by weight and now to be sold by volume.

There is a very strong argument for assembling the key data required for formulation, for all the raw materials currently in use or being considered, into a standard format readily accessible to the formulator. This may be a concise raw material data file set up on a computer or simply a card index system. Such data should not be cluttered with unnecessary facts but must include all key information required for formulation. The summary data card should, however, indicate sources from which more detailed information may be retrieved if required. A suggested format is shown in Figure 1.1

<table>
<tr><td colspan="3">R.M. Name:</td><td rowspan="2">R.M. Code Number:-</td></tr>
<tr><td colspan="3">Supplier:</td></tr>
<tr><td colspan="3"><u>Chemical Composition</u>:</td><td rowspan="2"><u>Hazard/Safety data</u>:</td></tr>
<tr><td colspan="3"><u>Physical Constants</u>:
Solids content:
S.G. of solution:
S.G. of solids:
Viscosity as supplied:
% Solids at 200 cP

Date:</td></tr>
<tr><td rowspan="2">COST:</td><td>per Kilo:</td><td></td><td>Full details in:-</td></tr>
<tr><td>per Litre</td><td></td><td>Filing Location:-</td></tr>
</table>

Figure 1.1 Summary data card

1.3 Clarifying the objective

It may appear to be stating the obvious to say that it is most important to clarify the objective of any project before starting work at the desk and particularly before starting any benchwork. To quote from the FSCT (Federation of Societies for Coatings Technology) introductory booklet, 'Today, in the paint industry, as in many other industries, progress is dependent on team work between production, sales, management and technical departments. All of these groups share in the creativity and productivity, which combine to make a successful company.'

Individual projects may be initiated at the request of marketing, sales, purchasing or production personnel or be technically generated but, whatever the origin, there will be an impact on all other functions if the project is successful. It is only logical therefore to ensure that the project has each of their input, if not wholehearted support, at an early stage rather than to find it blocked for a very good reason after completing a full development programme.

Preliminary discussions may reveal that the initial brief was not as clearly expressed as it might have been, or that it could create a difficult or impossible situation within manufacture if certain possible routes were followed. A further benefit of such discussion is that it helps to crystallise any thoughts on the logical and acceptable route to a satisfactory conclusion before leaping into action on the basis of a 'hunch' or 'first thoughts'. Possibly more important is that it helps to avoid the situation in which the target is impossible but the worker must be seen to be trying–such activity benefits no one and is best recognised before any time is wasted.

Having clarified the objective, it is good discipline and practice to state it clearly and unambiguously on a project initiation sheet. Cost, performance and anticipated completion date targets, if agreed, should all be recorded. Should a review of the progress of the project lead to agreement that the brief should be modified, such revised objectives should be recorded on the same initiation sheet. Figure 1.2 shows a possible simple project request form which acts as the project initiation sheet and to which all subsequent formulations and test results should be attached.

1.4 Identifying formulating constraints

Within the discussion to clarify the brief it is essential to identify any possible formulating constraints. These can arise particularly as a result of inadequate flexibility within the manufacturing operation. Progressive moves towards bulk handling and computer-controlled manufacturing operations, while giving positive improvements in productivity and hence a reduction in the cost of manufacture, can have the effect of restricting the opportunities for the introduction of additional bulk-handled raw materials or intermediate solutions. Problems arise from the lack of availability of segregated storage tanks, pumps and pipelines. A degree of flexibility may have been built into the system at the design stage, but may over the course of time, have become fully utilised.

Problems arising from such constraints can be avoided by formulating on existing standard bulk-handled raw materials. In the case of other key projects dependent on the use of a new non-standard raw material, discussion at the planning stage allows the production department to reconsider their operation and identify possible means of handling the new raw material by the time development work has been completed and bulk production is ready to start. These comments may seem obvious, but experience has shown that problems do arise as a result of lack of awareness of the potential problems and inadequate discussion at an early stage.

A further source of constraint is the proposed introduction of a raw material requiring special handling on account of its hazard rating, flammability or consistency. While means are available for handling materials in any of these categories, the introduction of such methods may be incompatible with the

TECHNICAL PROJECT REQUEST

To:	Date:
Project Title:	Project No:
	Category:

Objective:

Completion Required By: Signed: ....

Modification to original objective: Date:

Signed: ....

Date Project completed:

Hrs. Spent:

Report Issued: Signed: ....

Figure 1.2 Technical project request.

existing manufacturing operation or may require significant capital expenditure. In such circumstances the production department would have every right to refuse to handle them.

A third potential barrier to the introduction of a new raw material may be the purchasing department. Existing contracts will have been negotiated on the basis of anticipated offtake, and 'standard' costs established on those assumptions. A significant reduction in that offtake as a result of partial replacement by a material from another source could lead to an increase in the 'actual' cost, giving adverse cost variances against standard. Additionally, purchasing policy may have been established and agreed at board level, identifying preferred suppliers. Such opposition may be overcome by discussion, highlighting the benefits of the new raw material, but agreement to such introductions should always be obtained before starting a detailed evaluation.

Essentially this section reinforces the point that the technical department is part of a team and that successful innovation flows from a commitment of the whole company towards new opportunities. Time spent in evaluating new raw materials, however promising they may appear to be, will be wasted if subsequent events show that their introduction is not possible for one valid reason or another.

1.5 Use of standard formulation format

Formulations may be entered in a number of alternative ways, depending on the particular purpose for which they are to be used. In the simplest case the total composition is listed in terms of raw materials used with quantities expressed in appropriate units, depending on convention in each country. Commonly formulations were entered on a weight percentage basis, possibly because for many years paint was sold by weight and the importance of volume relationships had not been recognised. Today, however, paint is sold in most countries on a volume basis, either in litres or gallons and pints.

In such circumstances it is logical to formulate always to a constant volume. This makes true comparisons of the possible formulations, since variations in specific gravity of individual raw materials or the finished product will be taken into account. It will also be demonstrated in Chapter 4 that a number of volume relationships are important in paint formulation and such relationships can be calculated much more readily if formulations are always in volume terms. A well proven route is to calculate formulations to give a constant yield of 1000 litres, entering the quantities of individual raw materials in either kilogram or litre units depending on how they are normally dispensed. Trial laboratory makings of such formulations can use the same quantities, expressed in grams or cubic centimetres (cc), to give a yield of 1 litre – a quantity often appropriate to the completion of sufficient tests to evaluate performance.

If formulations are always entered in a common format and to a constant yield, later comparisons of one with another are simplified and recalculation avoided. A further benefit is that with experience the worker develops a 'feel' for what is likely to be a good starting point formulation for key ingredients of any new formulation. Since in many cases several minor variations of a basic formulation are made up at one time for comparative evaluation, it can be useful to use a formulation sheet which allows the entry of such variations, side by side, which reduces the amount of writing required and makes comparisons simple. A suitable form for basic formulations is shown in Figure 1.3. Note that a reminder to enter the project number, page number and date is at the top of the form and provision is made at the foot for relevant data such as viscosity, specific gravity etc. Such a simple format is adequate for ongoing work within the project, but is less ideal for costing purposes or for submitting a formulation to the factory for bulk manufacture. For these purposes alternative layouts are recommended.

FORMULATION SHEET

Paint Description:

Project No:
Page No:
Date:

Code No.	Ingredient	Paint Ref:-				
	Viscosity:					
	S.G.:					
	H.G.:					
	Signature:					

Figure 1.3 Formulation sheet.

Before leaving the standard form for entry of laboratory formulations, the way in which it is provided should be considered. Whilst there is no reason why the form should not be produced on loose sheets, there can be an advantage in the use of sequentially numbered duplicate 'laboratory formulation books'. The top copy, if perforated, can be removed for attachment to the ongoing project file, leaving a permanent record in the bound book which can be filed in the library, once completed. If formulation books and individual pages are pre-numbered there can be no possibility of confusion or accidental issue of the same formulation reference number to two different experiments. This may also overcome any last objection to the use of a 'loose-leaf' system for maintenance of the project file.

1.5.1 *Costing laboratory formulations*

Costing of formulations is most likely to be carried out by a simple computer program, where the print-out forms the record for the project file, but it can be done equally well manually. In either case, if the formulation is entered in a consistent structured way, a breakdown of the cost contribution of the various main classes of raw materials can be obtained, in addition to simplifying the calculation of the various key product constants such as pigment volume concentration (PVC), total solids content by weight and volume, specific gravity etc. (see Figure 1.4). Note that provision is made for grouping individual classes of raw materials such as pigments, extenders, resins, solvents and additives, allowing the total cost contribution of the individual classes to be shown in the right-hand column. In comparing formulations using mixtures of raw materials within each class, the key factors responsible for differences in cost can be very quickly identified.

It will be noted that the form also has space for inclusion of a handling loss allowance for individual raw materials or for the total yield. Such a column could be dispensed with if an alternative costing system used inflated prices for raw materials to allow for natural handling loss. The centre section of the form is included for entry of calculated figures for solid or volatile material from any solutions or latices used, on either a weight or a volume basis. From these columns the various constants referred to earlier can be derived quickly. The basis of such calculations is dealt with fully in Appendix I.

1.5.2 *Manufacturing formulations*

In entering formulations for use in a manufacturing operation, it is normal for the formulation to be broken down into stages, with the individual raw materials listed in the order in which they will be added during the making process. The composition of the individual stages will be determined by the nature of the product and the manufacturing method to be used. This aspect will be dealt with fully in Chapter 7. The concern in the context of this chapter is how best to indicate on one simple form the way in which the additions of the various raw materials are subdivided into the various stages.

A proposal for this is shown in Figure 1.5. It will be noted that in addition to giving the composition of individual stages, detailed notes are included to show how the additions should be made, times of mixing at each stage and any other information required to ensure that optimum manufacturing conditions are used and that the required standard has been reached before proceeding to the next stage. Space should also be allowed for the entry of

PRODUCT: | CODE NO: | STANDARD/ACTUAL COSTING
COLOUR: | FORMULATION DATE: | DATE:

	RAW MATERIAL	KILO	LITRE		WEIGHT		VOLUME		+	GROSS	STD/ACT	COST	TOTAL
					SOLID	VOLATILE	SOLID	VOLATILE	%	USAGE	PRICE		COST
				PIGMENT									
				EXTENDER									
				RESIN & OIL									
				SOLVENT									
				ADDITIVE									
				STAINER/ TINTER									

VISCOSITY:-	GROSS YIELD:-	1000	TOTAL:-	TOTAL COST:-
BS Cup:- Sec	LOSS ALLOWANCE:-		% SOLIDS (Wt) (Vol)	GROSS COST/LITRE:-
Krebs:- K.U.	NETT YIELD:-		S.G.:-	NETT COST/LITRE
Cone & Plate:Poise			P.V.C.:-	TINT COST/LITRE
Rotothinner: Poise	METHOD OF MANUFACTURE:-		P/b ratio:-	STANDARD/ACTUAL COST/LITRE:-

Figure 1.4

Laboratory Formulation

Ref: 123/456 Description: Green airdry paint Lab. ref: A 189
Document type: Revision PC ref: Date issued: 25/08/87
Reason for issue: Change of shading colour

	Signature	Date
Raised by	M. Smith	25/08/87
Keyed by	Eunice Jones	25/08/87
Authorised by	D.J.Brown	01/09/87

Safety regulations and special instructions

Contains lead pigment

Laboratory mixing instructions

Stage		Reference	Description	Premix	Mill	Kilos Make-up	Adjust	Summary
A1	Charge to high speed mixer and premix for 30 min	R1023	Alkyd 60%	149.0				674.0
		S0001	White spirit 100	99,6				119.0
		A0045	Calcium drier 10%	4.73				4.73
		P0102	Phthalo blue	13.4				13.4
		P0890	Iron oxide	22.4				22.4
		P0090	Mid chrome	47.3				47.3
		P0010	Titanium dioxide	5.98				5.98
B1	Pass through sandmill to heg. 7+							
C1	Wash mill and add to millbase with constant stirring	R1023	Alkyd (60%)		149.0			
D1	Add in given order to bulk whilst stirring	R1023	Alkyd (60%)			375.0		
		R0099	Hard maleic resin			82.2		82.2
		A0030	Lead drier 36%			6.97		6.97
		A0039	Cobalt drier 10%			2.49		2.49
		A0004	Methyl ethyl ketoxime			3.98		3.98
T1	Tint to shade	SC001	Shading colour white					
		SC006	Shading colour lampblack					
		SC008	Shading colour red oxide					
		SC012	Shading colour yellow oxide					
Q1	Adjust viscosity	452/000/9	White spirit 100				19.4	

Total kilos: 982.67
Total litres: 999.82

Figure 1.5 Laboratory formulation.

any deviations from the standard formulation and for the entry of the actual filled yield against the theoretical yield.

1.6 Use of test schedules

Having made a series of experimental paints, the next stage is to carry out selective tests, appropriate to the type of product being investigated and the basic project objective. Diligent recording of all results should be a matter of routine; they should be entered in their appropriate place at the time and not on scraps of paper or the sleeve of a laboratory coat for recording at some time in the future.

In line with earlier recommendations, the use of pre-printed forms for this recording of test results has much to commend it. For each class of product it may be necessary to have more than one form, depending on the depth of testing being carried out; indeed there is a very strong case for having a set of forms for each product group. In the initial stages of a product development programme a number of possible options, scattered around the area in which the final solution is anticipated, may be examined. As the investigation proceeds, some of the variables are progressively eliminated moving closer to the preferred formulation, with a final stage of 'fine tuning' the selected formulation to the optimum result attainable against the project brief.

In the early stages of the project, it is quite unnecessary and unproductive to test every sample produced to a full test specification. However, a number of key tests must be carried out to determine from those initial series of paints the direction in which the project should progress. As the project develops, so it will be necessary to test samples more fully to identify the preferred option. The final formulation selected should always be tested to a full product test specification to ensure that in moving to meet a new requirement or to achieve a new balance of properties, another key property has not been inadvertently upset. These considerations have led to the concept of having three stages of test schedule for each main class of product:

1. Preliminary test schedule.
2. Intermediate test schedule.
3. Final test schedule.

These staged schedules are complementary, with no duplication of tests. Forms can be used selectively at each stage of the project, depending on the basic objective and individual experiments.

One of the important benefits of such pre-printed test schedules, apart from simplifying recording, is that they act as a check list of tests which should normally be carried out. A further major benefit is that they simplify cross reference and review of earlier results. Typical product test schedule forms are shown in Figures 1.6–1.11. As with the standard laboratory formulation sheet,

PRODUCT TEST SCHEDULE

Product: WATER THINNED MATT EMULSION PAINTS

Project No:
Page No:
Date:

S. T. M.	TEST	Paint Ref:-				
	LIQUID PAINT PROPERTIES					
51-07-05	Condition of Paint					
	Settlement					
	Separation					
	Rusting					
	Structure					
51-06-01	Odour					
51-05-01	Specific Gravity					
51-01-01	Dispersion					
51-02-05	Viscosity Stability R.E.L.					
	Initial					
	1 week @ 43°C					
	2 weeks @ 43°C					
	4 weeks @ 43°C					
51-02-03	Viscosity Stability Krebs					
	Initial					
	1 week @ 43°C					
	2 weeks @ 43°C					
	4 weeks @ 43°C					
51-02-08	Structure/Gel Strength					
	Initial					
	1 week @ Room Temperature					
	2 weeks @ Room Temperature					
	4 weeks @ Room Temperature					
51-03-01	pH Stability					
	Initial					
	1 week @ 43°C					
	2 weeks @ 43°C					
	4 weeks @ 43°C					
T10/1						

Figure 1.6 Product test schedule – water-thinned matt emulsion paints.

PRODUCT TEST SCHEDULE

Product: **WATER THINNED MATT EMULSION PAINTS**

Project No:
Page No:
Date:

S. T. M.	TEST	Paint Ref:-				
	FILM FORMATION PROPERTIES					
52-12	Coalescence					
	a) Room Temperature					
	Plasterboard					
	b) Low Temperature					
	Plasterboard					
52-08-02	Colourant Acceptance					
52-08-04	Colour Uniformity					
52-12-02	Patching					
52-13-01	Patchiness on Tacboard					
	PAINT FILM PROPERTIES					
53-04-01	Opacity - Morest Chart 7K					
	Wet					
	Dry					
53-09-01	Adhesion					
	Initial (over old gloss)					
	After 1 week (over old					
	gloss)					
53-12-01	Mudcracking					
	Contrast Ratio (Computer)					
	Colour					
	a) Visual					
	b) Computer					
	PAINT FILM PERFORMANCE					
54-06-01	Water Spot Resistance					
	Mechanical Scrub-Gardner					
	Overnight					
	After 1 week					
T10/2						

Figure 1.7 Product test schedule – water-thinned matt emulsion paints.

PRODUCT TEST SCHEDULE

Product: **WATER THINNED MATT EMULSION PAINTS**

Project No:
Page No:
Date:

S. T. M.	TEST / Paint Ref:-				
	PAINT FILM PERFORMANCE				
54-07-02	Hand Scrub				
	Overnight				
	After 1 week				
54-09-01	Accelerated Weathering				
	Cracking				
	Blistering				
	Flaking				
	Chalking				
54-11-01	Exterior Exposure				
T10/3					

Figure 1.8 Product test schedule – water-thinned matt emulsion paint.

space is allowed for entering four sets of results in adjacent columns. It will also be noted that reference is made to the standard test method to be used, which can avoid confusion when more than one method exists appropriate to different types of products. As with the recording concepts discussed earlier, the use of product test schedules and their total completion during a development programme should never be mandatory. The forms are there to be used selectively and sensibly. Experience has shown that, given such flexibility in use, most technologists prefer to employ them, rather than using blank sheets of paper or a laboratory notebook.

While the examples shown refer to decorative and trade sales products, the concept can equally be applied to industrial product development. Often such products are designed to meet a particular specification which identifies the range of tests and methods to be used in the evaluation of experimental products. Similarly, product test schedules could be specific to individual customers, using them as a checklist on application conditions and for schedules and important performance requirements. The concept is proven – the value in any other development environment will depend on the correct design of form and selection of tests to meet the specific needs of that environment.

PRODUCT TEST SCHEDULE

Product: SOLVENT THINNED GLOSS FINISHES

Project No:
Page No:
Date:

S. T. M.	TEST	Paint Ref:-				
	LIQUID PAINT PROPERTIES					
51-07-05	Condition of Paint					
	Settlement					
	Synaresis					
	Skinning					
51-06-01	Odour					
51-01-01	Dispersion - H.G.					
51-05-01	S.G. (± 0.02)					
51-04	Non-Volatile Content					
	Viscosity Characteristics					
51-02-03	Kreb Stormer/25°C					
	Initial					
	1 week @ 43°C					
	2 weeks @ 43°C					
	4 weeks @ 43°C					
51-02-05	ICI Cone & Plate					
	Initial					
	1 week @ 43°C					
	2 weeks @ 43°C					
	4 weeks @ 43°C					
	Ferranti Shirley Plot					
	1,000 sec^{-1}					
	10,000 sec-1					
	Structure					
	Visual					
51-02-08	ICI/Sheen Gel Strength					
	Overnight					
	1 week					
T18/1						

Figure 1.9 Product test schedule – solvent-thinned gloss finishes.

PRODUCT TEST SCHEDULE

Product: SOLVENT THINNED GLOSS FINISHES

Project No:
Page No:
Date:

S. T. M.	TEST	Paint Ref:-				
	FILM FORMATION PROPERTIES					
52-04-1	Wet Edge					
52-06-	Sagging - Sheen Sag					
	Applicator (50μ - 275μ)					
	Colour Development -					
	Flooding & Flotation					
52-09-01	Rivelling					
	Drying Properties BK					
	Recorder					
52-11-01	a) Drying at Room Temp.					
	1. Before Incubation					
	2. After Incubation					
52-11-04	b) Drying - Low Temp.					
	40°F/50% R.H. - BK					
	Recorder					
	1. Before Incubation					
	2. After Incubation					
	PAINT FILM PROPERTIES					
53-01-01	Colour					
	a) Visual Comparison					
	b) Computer					
53-02-02	Gloss (60° Meter & 20°					
53-02-04	Meter)					
53-03-01	Flow					
53-03-02	Flow from Sharp Edges					
53-04-01	Opacity - Morest Chart					
	using "K" Coater (No. 7)					
	a) Wet					
	b) Dry					
T18/2						

Figure 1.10 Product test schedule – solvent-thinned gloss finishes.

Product: **PRODUCT TEST SCHEDULE**
SOLVENT THINNED GLOSS FINISHES

Project No:
Page No:
Date:

S. T. M.	TEST Paint Ref:-				
	PAINT FILM PROPERTIES				
53-04-02	Opacity - Contrast Ratio				
	using Film Applicator				
53-09-01	Adhesion - Cross Hatch				
	Test				
	a) Initial (over				
	b) After 1 week (over				
54-02-01	Yellowing (Woburn)				
54-09	Weathering - Accelerated				
	(Weatherometer)				
	N.B. The following tests				
	are required for Non Drip/				
	Gel Paints				
51-02-10	Stevens Gel Strength				
	Overnight				
	1 week				
	1 month				
51-07-08	Accelerated Storage				
51-07-09	Shop Window Stability				
52-03-07	Application				
	Gloss Holdout & Adhesion				
	over M.A.S.P.				
T18 3					

Figure 1.11 Product test schedule – solvent-thinned gloss finishes.

1.7 Understanding the costing system

Constant reference is made throughout this work to the need for competitive formulations meeting both performance and cost targets. To be cost effective, the technologist must understand how the company's costing system is applied and the various factors which can influence cost, possibly increasing it to a figure unacceptable to sales or marketing. Potential problems can often be overcome by discussion with financial or production colleagues and by highlighting areas which may be a barrier to achieving desired results.

It is not suggested that chemists should become cost accountants, but it is important to recognise the various influences and to identify formulation routes or other changes which could lead to more cost effective manufacture.

Costs normally contributing directly to the overall finished product cost, often known as the FFC or full factory cost, are:

1. Raw material cost.
2. Manufacturing cost.
3. Container and filling costs.

Each of these costs can be influenced by management decisions on the allocation of various fixed costs or by other factors such as scale or method of manufacture, handling methods, diversity of raw materials, product range etc. Principles of allocation of costs are normally based on historical records or established patterns of operation and rarely have sufficient flexibility to adapt to a new situation in the short term. However, by understanding the existing system, it is possible to indicate where changes may be necessary to adapt to new cost-saving opportunities. These various costs will now be examined in more detail.

1.7.1 *Raw material costs*

This may appear to be quite unambiguous. However, as a result of different approaches to cost accounting the formulator may be faced with 'actual' cost, which is equivalent to the current buying price, or with 'standard' cost, which is a projected average buying price for individual raw materials established at the start of the financial year as part of the budgeting process, and used as the basis of 'standard' product costs. Use of 'standard' costing avoids the problems created by price variations in raw materials during the year; any differences between 'actual' and 'standard' are recorded as positive or negative 'variances' within the ongoing budgetary control procedures. From the formulator's viewpoint the important point is to be aware of these differences and to ensure that like is compared with like when assessing the relative costs of two formulations, particularly when comparing an experimental modification with the existing standard formulation.

Whatever basis of costing is used, allowance is normally made for anticipated loss in storage or handling, normally expressed as a percentage which may range from 0.5–5.0% depending on the nature of the raw material and the way in which it is handled within the manufacturing operation. Provision for this may be made by inflating the 'actual' or 'standard' cost or by inflating the total cost of the individual raw material *usage* within the batch. Established loss allowances may become unrealistic if alternative methods of handling, for example, changing from bag deliveries to bulk or drums to tankers, within a move to partial or full bulk handling are contemplated. Similarly, a reduction in the diversity of raw materials used across a range of formulations can increase the rate of turnover of individual raw materials, with consequent reduction in loss or 'write-off'.

Loss allowances can also escalate in their impact on total formulation cost when formulating and manufacturing on 'intermediate solutions'. Handling losses may be applied to the raw materials used in the manufacture of the solution, and to the completed solution which is held in bulk for use in finished product manufacture. A further loss may be applied to this solution as a raw material used in the full batch. This is a case where the duplication should be resolved by dialogue, once it is recognised that it is occurring.

1.7.2 *Manufacturing costs*

These costs include the allocation of total manufacturing, labour and overhead costs to a particular method of manufacture, dependent on batch size, time of utilisation of the equipment used, and the relative use of labour, energy etc. within the process. Cost allocations should be reviewed when major changes in operating practice arise. One of the most significant factors to influence manufacturing cost is batch size – moving to much larger batches or to a continuous process can dramatically reduce costs. It may be thought that the formulator can have very little influence on this, but this is not so. A range of related products may be in regular manufacture in relatively small batches. Analysis of their composition may establish that all could be produced from a common stabilised mill base, with the individual characteristics being achieved by differences in the let down and completion stages. Use of such 'master batches' is a very positive route to lower cost manufacture. Similarly, in the production of a range of 'ready mixed' colours, inevitably some tend to be slow moving and only justify small batches. These are usually at the 'deeper' end of the colour range, calling for dispersion of dry pigment rather than post tinting. An alternative method of manufacture is the production of a limited range of concentrated base colours, from which individual colours can be produced by blending. Both of these examples increase the scale and reduce the frequency of production of the dispersion stage, which normally carries higher charges than subsequent stages of manufacture.

Another potentially high cost area is the tinting stage, particularly if carried out manually, which increases in direct proportion to reduction in batch size.

Below a given size, manual tinting is quite uneconomic and consideration should be given to use of accurately dispensed controlled strength tinters.

1.7.3 *Container and filling costs*

These costs can be interactive and an apparent reduction in one can be offset by an increase in the other. Filling costs per litre increase with reduction in container size or with a multiplicity of filling sizes. Change to an alternative design container may affect the speed of the filling operation and influence cost. Moving from metal to plastic containers may affect container costs positively or negatively, depending on size, but may bring in a consequential cost increase as a result of a higher level or different type of in-can preservative being required. Normally decisions on change in these areas are made by marketing or production management but there should be a technical involvement to ensure that potential problems requiring minor modification of the formulation are not overlooked. Certainly any proposed pack change should be fully evaluated by the technical department before adoption.

1.8 Know the competition

In any competitive industry – and the paint industry is no exception – it is important to maintain a detailed knowledge of how the company's product compares with that offered by main competitors. In a situation where an established product has areas of benefit over the competition it is important to make marketing and sales colleagues aware of those differences so that they may choose to exploit them by aggressive marketing. They may, however, judge that the market is not very receptive to such 'benefits' and that cost is the determining factor in which product is chosen within a broadly related level of performance. In such a situation the product benefit is contributing nothing directly to sales or market share and could be regarded as an area of overformulation, leading to a potential opportunity for cost saving.

Markets tend to be cyclic in their response to cost and quality, and a dynamic team will react to these opportunities for increased market share by adjusting the balance of properties to suit the situation of the time. It should always be remembered that there is no such thing as the perfect paint; formulation is always a compromise and every manufacturer will tend to adopt or achieve a slightly different balance of properties or compromise, depending on the preferred range of raw materials used, methods of manufacture and approach to cost and profit. Within any decision to change, or not to change, the company must know how it stands relative to the competition. It is the technical department's responsibility to make the technical appraisals and to ensure that the marketing department is regularly updated.

Such comparative testing should be structured on a formal basis, using a monitoring programme which examines different classes of products within a planned sequence. The programme should be such that each class of product is examined at least once a year – in some circumstances more frequent examinations may be advisable. Evaluation should be kept as simple as possible, but should preferably use objective tests allowing results to be recorded and filed for cross reference at succeeding evaluations. Over a period a pattern of the way the market and individual competitors are moving and whether the company is keeping in step is built up. An incidental benefit of such regular monitoring is that it may occasionally pick up an important change in a competitor's product before such changes are promoted in a new marketing campaign. It is quite common for such changes to be fed into the distribution chain prior to introduction of a new pack design which extols the 'new improved' quality. Advance warning of such activity can give the marketing department valuable time in planning a possible response.

In examining retail or trade products, tests may be restricted to measurements of viscosity, total solids contents, specific gravity, draw down at constant film thickness on black and white contrast charts, measurement of gloss or sheen and a simple test for film strength – in the case of latex paints possibly a simple wash or scrub resistance. Selective tests such as stain resistance or removal, burnish resistance, accelerated weathering, permeability, application characteristics etc., may be added to the basic list according to the nature of the products being evaluated and their performance requirements. This is clearly another area which lends itself to the design of simple, clear forms for the recording of results to act both as a checklist of which tests should be carried out on different type products, and a means of making direct comparisons on future occasions.

2 Establishing a structured development programme

RICHARD WOODBRIDGE

2.1 Introduction

One of the fascinations of working in the paint industry is that there are always challenges which act as stimulants for any dedicated paint chemist. The competitiveness of the industry means that the chemist must be constantly aware of the cost implications of any change and the often ridiculously short time scale in which an answer is required. Yet the chemist must know that at the end of the line is a discerning user who will very quickly complain if the product does not come up to expectations. On the one hand, it is felt that short cuts are the only way to tackle the problem, while scientific training demands a disciplined systematic approach with extensive testing before releasing the formulation to the factory.

Interesting differences apply between formulating a product for use in industrial finishing and a decorative product for retail distribution. In the first case it may be argued that an important feature is the product's handling characteristics on the production line, whether it be spray, dip, flow coat or whatever. This suggests a limited trial run in the user's factory as early as possible in the development programme.

In the case of the retail decorative product, one of the important features is shelf life, which may be anything up to three years, or more in the case of slow moving ready mixed colours. Whatever the conditions of storage the product is expected to behave normally when eventually used. Forecast of storage life is important and the longer the time that can be devoted to testing, the more ultimate results can be predicted with confidence.

These are just two obvious examples of the many differences which exist within product development – it is true to say that every project is different and requires an individual approach. However, some broad guidelines follow for different types of project which should be sufficient to identify the various considerations that must be taken into account at the planning stage of any project.

2.2 Raw material substitution

This may appear to be the simplest reformulation task facing the chemist and in many cases, particularly if the raw material in question is regarded as a 'commodity chemical', it may be argued that no laboratory testing is necessary.

Yet what can be regarded as a commodity chemical in the paint industry? Does one grade of white spirit have the same distillation range, aromatic content and solvency as another? Does one sample of whiting have the same particle size distribution, oil absorption and colour as another? Do two long-oil alkyd resins of similar oil length, general composition and solids/viscosity relationship have similar acid and hydroxyl values? At what point can one say that the differences are so small that they are not important? Ultimately it is experience that will tell what risks to take in making a substitution with little or no testing, but experience can be gained the hard way and it is usually wise to be cautious in giving alternative approvals.

It is clearly better to avoid having to make snap judgements. Working closely with the buyer or purchasing department, it is a good discipline to identify two or three preferred suppliers and to evaluate and approve 'equivalent' grades of key raw materials. With such approvals, the occasions when snap judgements are required become more limited.

However, even here there is a word of caution. The case for formulating the total company product from as few raw materials as possible has been argued earlier. This inevitably means that individual raw materials are used in a number of different products. It is by no means uncommon for an alternative raw material to be perfectly satisfactory in some products and not in others, leading to restricted approvals of the alternative.

The decision on the level of testing required must therefore depend both on the type of raw material being substituted and the criticality of the end product. At the very least it may be advisable to carry out some simple comparative tests, such as colour, oil absorption, distillation range etc., depending on the class of raw material. Even better would be to evaluate the two alternatives in a typical formulation, testing selectively for those properties which might be influenced by subtle differences.

The above assumes that, to all intents and purposes, the raw material being considered is a direct equivalent of the previously used material. There are occasions, however, when an alternative which is not directly equivalent may be considered. The question then is to decide on what basis the existing material should be substituted. Should the replacement be made on an equal weight basis, or on equal volume, at equal cost, or indeed on any other criterion? All approaches can be justified in particular circumstances – the important action is to think critically of the function of the material being replaced and what is required of the alternative to provide the same function. To illustrate this point several examples will be considered.

1. An alternative fungicide has been offered, claiming improved performance over the currently used grade. In this context it is assumed that the aim is to maintain existing performance, which is judged to be perfectly adequate.

A good approach would be to make a trial laboratory sample of a representative product, eliminating the standard fungicide, and add to representative samples:

(a) the standard fungicide at standard level;
(b) the alternative fungicide at equal cost.

Sample (b) would then be blended with the untreated paint to give 50%, 70% and 85% of the standard fungicide cost. All paints can then be evaluated for fungicidal properties to establish which level of the new material gives equivalent performance to standard. Having established the optimum level and identified a potential cost saving, the modified paint should be tested fully to ensure that the new raw material has not affected other paint properties. In this particular case it would be wise to include tests for effect on colour and odour during storage.

2. For rationalisation purposes, it has been decided to standardise on one long oil alkyd rather than have three variations of slightly different viscosities. Products based on the extreme viscosity grades will be reformulated to use the existing standard intermediate grade. In this case the determining factor will be to ensure that the standard viscosity of the finished product is maintained. Hence some reduction in resin content and overall total solids will be required in the products previously formulated on the lower viscosity alkyd and, conversely, those previously on the high viscosity resin will require an increase in resin content and total solids.

However, since the viscosity of the resin may influence dispersion efficiency, the chemist should carefully examine the influence of the change on the contrast ratio of the various products and some adjustment of TiO_2 levels may be required. If some thixotropic resin is normally included in the formulation for rheology control, it may be necessary to make a slight change in the ratio to maintain the desired properties. Finally, the overall effect of changes necessary to maintain performance should be costed to establish whether the savings achieved by rationalisation have been offset by the other changes necessary.

2.3 Cost reduction

From time to time, particularly when the market gets tough, the formulator will find that cost reduction exercises dominate activity, since marketing and sales colleagues will argue that it is the cost of the product which is the main reason for lack of sales. In such exercises, it is rarely accepted that quality can be lowered in line with reduced costs. The role of the formulator is to ensure that the required quality is formulated in the most cost-effective way possible.

This immediately raises the question of whether the product is over-formulated and is achieving a higher standard than that expected by the customer or offered by the competition. A critical evaluation against the specification of the competitive product should identify any areas where performance may be sacrificed and the revised target should be clearly noted and agreed before commencement of the reformulation.

A further recommended discipline when reconsidering any formulation, and particularly in the context of a cost-reduction exercise, is to examine it carefully for every possibility of simplification. Formulations have a tendency to 'grow' over the years, with a little of this or that added to overcome a problem or to boost a particular property. Cases have been known where extremely complex formulations became established as a result of a chemist, working to a tight deadline, finding that a blend of two existing products just happened to meet a new requirement. The quick answer was to enter the blended formulation, on a calculated individual raw material basis, and continuing work pressures never allowed time for simplification. This may be an extreme case, but it has occurred.

More frequently it is found that minor additives are included because 'we always use them' or 'experience has shown that they give a benefit'. That view may well have been justified in the past, using a different resin system, but that is no justification for continuing use when developments in the resin or solvent system have eliminated the need. Even changes in market requirements or competitive standards may make additives redundant whereas they could have been essential when first introduced.

Typical of the questions which should be asked when re-examining a formulation are:

- Can a blend of three or more resins, solvents or extenders be justified to achieve the right balance of properties?
- Can rheology control be achieved by more careful selection of the resin system or in choice of extender, rather than by relying on structuring additives or anti-sag agents?
- Could an alternative dispersing agent be equally effective but eliminate or reduce the need for defoaming agents?
- Is it necessary to continue using an 'in-can' corrosion inhibitor when the product is now distributed solely in plastic containers?

It is also a wise precaution to review all other costs allocated to the total cost of the product before assuming that reformulation must be undertaken. These costs, over and above the cost of raw materials, add up to a significant contribution to the 'full factory cost' and more realistic costing may make all the difference between success and failure. Areas to question are:

- Is the lowest cost container, appropriate to the product, being used and is it correctly costed?

- Are the 'loss allowances' applied to the various stages, e.g.
  – raw material usage
  – handling of intermediate solutions
  – batch manufacture
  – filling
  realistic and appropriate, taking batch size and method of manufacture into account?
- Do the manufacturing, labour and overhead costs relate to the true cost of manufacture of the particular product?
- Are the filling costs realistic?

It may be assumed that all such cost allocations are sacrosanct and not to be challenged, but in a situation where cost reduction of a particular product is essential to obtain market share or new business, no area should be ignored. Only by scrutinising every factor influencing cost will the most competitive position be established.

Having completed these essential preliminaries, it may be accepted that further cost reductions can only be achieved by reformulation. A starting point is to look logically at those items which contribute most to the overall raw material cost – in other words by reviewing the formulation, costed out fully. Normally it will be found that the two main contributors to cost are the pigmentation and the resin. Before considering these individually we should recognise that total RM (raw material) cost is governed by total solids content. Is it possible to reduce this by 5–10% without affecting application characteristics? Use of a slightly higher viscosity grade of resin or, in the case of emulsion finishes, a higher molecular weight colloid may achieve the desired reduction without upsetting the general balance of properties. The advantage of this approach is that it reduces all ingredients proportionately and hence has minimum effect on any individual performance property. Such a small reduction in both the pigmentation and overall solids is likely to have a less significant effect on opacity, build or performance than achieving an equivalent cost reduction by reduction or substitution of a single component.

Looking next at the resin system, it should be questioned whether the most economical system is being used to achieve the existing or revised standard of performance. This does not necessarily mean using the lowest cost resin available. Reference has been made in Chapter 1 to the influence of bulk purchase or reduction in diversity of raw materials on costs. It is also possible that use of a slightly higher unit cost resin, of higher performance, can allow a higher pigment volume concentration (PVC) or, in other words, a reduction in total resin content.

In considering pigmentation, particularly in white and pastel finishes, titanium dioxide (TiO_2) level is determined by the desired level of opacity at the required spreading rate. What scope there is for reducing the level depends on the degree of dispersion being achieved and maintained during the drying process i.e. the efficiency of utilisation of the TiO_2 employed. This is possibly one of the most rewarding areas for study. Independent reviews by major TiO_2 producers have shown very significant differences in utilisation between equivalent products available in the market-place. It should be every paint formulator's objective to ensure that the product is in the upper quartile of results from such comparisons and league tables.

Areas to study are the optimisation of wetting during the dispersion process and the avoidance of flocculation during the completion stages or during application and drying. In the case of solvent-thinned finishes, dispersion in a low viscosity resin, possibly with higher than normal hydroxyl or acid value, may give an opacity improvement equivalent to an additional 5–10% of TiO_2. In emulsion-based products, use of more efficient pigment wetting and dispersing aids and avoidance of high viscosity solutions during the dispersion stage, can achieve significant reductions in TiO_2 without sacrificing opacity. Some improvement may also be achieved by a slight increase in PVC, introducing a degree of 'dry hiding' into the product. This aspect is dealt with in depth in Chapter 4.

In coloured finishes, the choice of pigmentation is critical to the achievement of the required colour at the specified level of opacity, durability and performance. Inclusion of the maximum percentage of inorganic pigment such as yellow and red oxides of iron is desirable from all points of view. Selection of the correct grades is clearly important – experience suggests use of the cleanest, yellowest tones in both red and yellow iron oxides can reduce the need for more expensive, less opaque and less durable organic pigments. Optimisation of the pigmentation for any colour, taking into account the required level of opacity at minimum cost, can best be achieved by one or other of the various instrumental colour matching systems available today, programmed for the standard range of pigments used in a particular range of products. Their cost, whether purchased outright or leased, can usually be fully justified by the savings achieved from intelligent use of their full potential.

2.4 Product improvement

This area of development may utilise a wide range of techniques, depending on the scale and nature of the improvement sought, ranging from a slight uplifting of a key property, such as opacity, drying rate, corrosion resistance etc., to a more major reformulation to upgrade general performance or durability. Even an apparently simple change may call on all the skills of the formulator and an extensive development programme to achieve the new level of performance, particularly if adverse cost penalties are to be avoided.

Assuming that the existing formulation is cost effective and has eliminated any area of unnecessary cost, it should be remembered that there is always a price to pay. Raising one property is likely to increase total raw material cost unless it is possible to sacrifice another property to compensate, i.e. to readjust the balance of properties and to achieve a different compromise. It is wise to recognise this implication right from the start and to agree cost targets and any change in specification with either sales or marketing colleagues or the customer before starting any programme of development work.

Within the potential diversity of attack, depending on the nature of the improvement sought, there can be no overriding rules on the 'right' approach to use but some basic principles still apply. One of the first things to consider is whether re-inventing the wheel can be avoided. Existing records may show that an alternative formulation exists which more closely matches the new requirements, either within established products or possibly in formulations investigated during the original development of the product. This highlights the importance of being able to compare existing formulations for cost and performance against key parameters using existing data. The alternative would mean having to remake experimental quantities and retest to establish differences. Such 'desk research' can only be done if good records are established and maintained, as proposed in Chapter 1.

Equally, the chemist should never be ashamed of using expertise and formulations from other sources to help solve a problem – there will always be another unique problem to solve in the future. Always take whatever help can be obtained from any source to reduce the amount of work necessary to solve the problem. It should be remembered that bonus points are earned for solving problems quickly and efficiently, not for the amount of work undertaken to achieve it. Having completed these preliminary steps, the design of the project and the best way to achieve the desired improvement can be tackled.

Assuming that no obvious 'short cut' has been identified, the existing formulation should first be critically reviewed to identify those raw materials which are unlikely to have any significant influence on the selected area for improvement. In the early stages of development these may safely be left unchanged. This helps to identify those areas of the formulation which must be reconsidered to achieve the new performance. Ideally, the number of variables identified for investigation should be reduced to the absolute minimum to simplify the programme undertaken. Depending on the anticipated scale of the project, a decision can be taken on whether to tackle it in a stepwise manner or on a more statistical basis, varying several components simultaneously, but in a way in which results can be analysed to establish the key influences.

Within this decision on approach, the possible interactions which may occur must be considered and allowed for within the programme. Investigation of an alternative key raw material, such as resin or pigment, to achieve the desired improvement may require a change in solvent balance or in pigment dispersing aid. Careful study of the supplier's literature should identify whether such associated changes should be made, or at least considered, at a later stage within a fine-tuning exercise. Within this decision the difference between changes which are essential, those which are desirable and those which are not important and only exist because of unrelated factors must be identified. As stressed in Chapter 1, change for change's sake will only proliferate stock inventory and add to costs.

In the simplest case of product improvement it may be necessary only to increase the content of a particular raw material, to change a ratio of usage of two components, or to substitute an existing raw material with one of higher performance. Due note should be taken of the effect of such change on overall product parameters such as PVC, total solids and volume percentage, and compensations made accordingly. Small laboratory makings, evaluated for essential properties, can then establish whether the desired improvement has been achieved without sacrifice of other properties. Comparisons should always be made against the existing standard formulation, made on the same scale and under similar conditions. Invariably differences exist between laboratory and full scale works manufacture which must be allowed for at the time of scale up (see Chapter 7). Failure to recognise this within comparative tests could lead to incorrect conclusions being drawn.

In evaluating the effect of change at this early stage of development, it is important to restrict the level of testing to that necessary to show that the key properties are in line with the new specification. In other words, use of a simple 'first stage' test schedule, as recommended in Chapter 1, is all that is required and can help to identify more quickly whether the prime objective has been achieved. Clearly the selection of the range of key tests depends on the product and the improvement sought, but should always include a few basics such as viscosity characteristics, drying and surface finish (level of gloss, sheen, etc.) to ensure that basic properties have not been inadvertently changed. The logging of these test results against the formulation record adds to the background knowledge accumulated over a period of time and may save considerable time in a future project.

Once progressive developments and testing have indicated that the main objectives have been achieved, it is important to carry out a 'full test schedule' to ensure that the product meets *all* requirements expected of it. As a result of this more extensive testing it may be found that a small amount of fine tuning of the formulation is necessary to bring all properties into balance.

The final stage of development is to scale up and produce a trial works making, simulating as far as possible the manufacturing techniques to be used in bulk manufacture. This trial batch should be tested against both the final experimental laboratory making and a standard works batch of the existing formulation. This aspect is covered more fully in Chapter 7.

Much of the above applies equally to the development of a more significant improvement of an existing product, but in such cases further fundamental changes may be required at the outset. In the extreme case the approach required would be the same as that needed for the development of a new product, as discussed in Section 2.6. Every project is likely to be different, and for the formulator the challenge is to identify at a very early stage which route to reformulation is most likely to achieve the desired result in the shortest possible time. Ultimately experience, supported by existing records, will assist in identifying the best approach to be taken.

2.5 Matching the competition

In Section 1.8 the importance of knowing exactly how the product stands compared with the main competition has been stressed. In a static situation it is likely that the main competing brands will settle down to a roughly uniform level of performance, although more critical comparisons may show subtle differences in balance of properties – possibly significant to the paint technologist but not easily recognised by the customer.

Unfortunately the market rarely remains static, and a company is often faced with the situation that a competitor makes a change which gives them a commercial advantage – either by claiming a product benefit or by offering a lower price. In such a situation, failure to react within a reasonable time will lead to loss of market share. The technologist is then faced with an urgent programme to produce a match as quickly as possible.

The nature of the project undertaken depends on the difference identified and may follow the lines suggested in Section 2.3 for cost reduction or in Section 2.4 for product improvement. Whatever the change, it is important to avoid the trap of overreacting and immediately engaging in a major reformulation when possibly all that is required are slight changes.

It is also important to recognise that it is most unlikely that two companies will ever produce exactly the same product. Because of slight differences in manufacturing conditions, two factories producing to exactly the same formulation are likely to achieve products showing subtle differences. In other words, it is necessary to establish what the significant differences are and to accept that an exact match is both unlikely to be achieved and unnecessary.

The technologist must also avoid the trap of bringing complexity into the raw material inventory by automatically assuming that a change in resin, or other key ingredient, is necessary to achieve a match. Whilst in some cases this may be the only means of achieving a similar product, the possibility of achieving an equivalent product using existing standard raw materials should have first been eliminated. Complexity invariably adds to cost and in this competitive situation cost effectiveness is paramount.

Another point to stress is that it is normally only necessary to match the competitive product in application and performance, and not in other characteristics which are of lesser significance. As an example of this, consider the case of a decorative interior flatwall finish for retail sale in the DIY market. The customer is concerned with ease of application and the ability to achieve a good uniform opaque finish in two coats using brush or roller. Depending on the price sector, the customer may have certain expectations of performance in terms of mar resistance and cleanability. One product may be formulated at relatively low brushing (high shear) viscosity but require a fairly high loading of TiO_2 to achieve the necessary level of opacity. An alternative product may be formulated at slightly higher 'high shear' viscosity and require a lower level of TiO_2 for equivalent opacity at normal spreading rate. The second product will give a slightly lower spreading rate but this difference is unlikely to be significant to the occasional DIY user. In practice more paint than is actually required for the job is purchased and the residue sits in the garage or store until eventually discarded.

Such an approach would be quite unacceptable in a product required for use by the professional tradesman, who knows exactly how much paint is required for a particular job and would very quickly recognise a reduction in spreading rate necessary to achieve an acceptable level of opacity.

Taking an alternative example, in the industrial finishing sector, the first requirement is that the product can be applied under conditions in use in the factory to give a film free from defects, which cures fully under the standard curing conditions. Whilst it may be possible to persuade the customer to change the application and curing conditions to suit the alternative product, this will always be an uphill battle and puts the alternative product at an immediate disadvantage. Satisfying these conditions is usually of greater importance than subtle differences in ultimate performance.

In considering what testing should be carried out, it is important to recognise these differences. The technologist must assess which are the most important characteristics to the end user and identify which tests are most likely to establish whether these vital characteristics have been achieved. Failure to recognise this will inevitably lead to the product being rejected at an early stage.

2.6 New products

From time to time a chemist is asked, or sees the opportunity, to formulate a new product. It should first be questioned what is meant by a 'new product'.

- Is it a product new to the company, but similar to that already offered by a competitor or in a different market? In such cases, the guidelines to follow are those given in Section 2.5.
- Is it a development of an existing product, coming into the category of product improvement (see Section 2.4)?
- Is it a lower cost version of an existing product, where the guidelines of Section 2.3 apply?

Once again, it is found that the basic questions are being asked before leaping into action.

It may be decided, after careful review of available knowledge has failed to identify any logical starting point, that new ground is indeed being broken. Alternatively, the concept may have been triggered by the availability of some new technology offered by a raw material supplier. In this case the guidelines in Section 2.7 on how to avoid the 'NIH' ('not invented here') factor and how to use all the information available as a basis for the development of something unique to the technologist's own company should be studied.

When it is finally decided that the concept really is new, the technologist must start from first principles.

First the chemist must set out, as far as possible, the product specification in very broad compositional, performance and cost terms and obtain agreement of these from marketing, purchasing and production colleagues. There is no point in undertaking an 'in depth' programme of development only to find that there is, in their view, no market, no means of producing it, or it uses raw materials unacceptable for one reason or another. Early involvement of such colleagues normally ensures their interest, and readiness to support the project at a later date.

The setting up of broad specifications should be considered in more detail. First the composition, in terms of raw materials likely to be considered for investigation, should be studied, and the options against each category of raw material listed (see Table 2.1). This list will obviously vary according to the nature of the product being developed. Within the grades of raw materials listed for investigation, those currently stocked as 'standard' should be included wherever possible.

Table 2.1

Category of raw materials		Grades to consider
Pigment	– TiO_2	
	– Coloured	
	– Extender	
	– Functional	
Resin	– Prime	
	– Modifying	
Solvent	– Prime	
	– Diluent	
	– Other	
Additives	– Dispersing aid	
	– Colloid	
	– Bactericide	
	– Catalyst	
	– Rheology modifier	
	– etc.	

Next consider the formulation from a broad quantitative viewpoint, listing the possible range of key ratios or constants, such as:

Total weight solids	e.g. 50–60%
Total volume solids	e.g. 30–45%
PVC or p/b ratio	e.g. 40–50% or 0.9–1.1 : 1
Likely specific gravity	e.g. ~ 1.5

Range quantities for the individual raw materials previously listed can now be entered using the key ratios or constants as a check that quantities are kept within realistic levels. Guidelines on typical levels of representative raw materials are given in Chapter 3.

Calculation of the mean composition within the range identified will give a possible starting point formulation while the limits will indicate the extreme realistic position. Costing the 'mean' and 'extreme' formulations will give indicative costs of the new product.

Finally, the specification should include anticipated performance data against key test areas, for example

Rheology
Drying/curing conditions
Finish – e.g. gloss at 20°/ 45°/ 60°
Hardness
Contrast ratio
Colour
Special requirements

It may not be possible to enter figures against all of the above, but the more the technologist tries to complete the picture before making the first experimental sample, the better the overall picture will be of the way the formulation is likely to develop.

Having reached agreement in principle to the concept, the development programme can then be started. How this is tackled will depend, to a certain extent, on the number of variables identified in the first stage, on the nature of the product being developed and the critical nature of the performance targets to be achieved. Options include:

- A stepwise approach, resolving as far as possible particular aspects of the formulation at each step.
- Varying more than one item at a time, but using traditional statistical design of the experiments to facilitate analysis of the results.
- A multivariable approach, in line with one of the computer techniques discussed in Chapter 6.

There is clearly no one preferred route to recommend since every project will be different. The following examples may help to illustrate the broad approach and suggest a logical route for other projects.

2.6.1 *Project 1.* To develop a high-build semi-gloss stoving finish for spray application directly on galvanised steel

Two key areas to establish are: (i) the resin system which meets the specified stoving conditions, gives adequate build at expected spreading rates and good adhesion to galvanised steel; and (ii) the pigmentation required to give the required colour, opacity, sheen and corrosion resistance.

It may help in this case to first establish in a 'simple' medium the prime pigmentation required to give the required colour and opacity, taking into account possible contributions to colour from any anticorrosive pigments being investigated. Clearly, choice of the latter may be restricted by the colour required for the finish. Alternative extenders and anticorrosive pigments which give approximately the right sheen and general finish can then be examined.

Using this 'basic' pigmentation, the alternative resin systems can then be examined in more detail to identify the preferred system for achieving the desired build, adhesion and general film characteristics.

Having narrowed the choice of pigmentation and resin system, a range of paints can be prepared, varying ratios within those predicted at the first stage, to test more fully for adhesion, corrosion resistance, durability and full physical characteristics.

During these various experimental stages the other aspects of the formulation, such as solvent balance, need for rheology modifiers, etc., will have been fine-tuned.

It might be argued that this particular example lends itself more to a multivariable approach from the beginning, but it does illustrate how the project can be simplified using a stepwise approach, leaving the more statistical methods to the later fine-tuning stage of the formulation.

2.6.2 *Project 2*. To develop an aqueous 'antispatter' eggshell, tint-base finish for walls with film properties equivalent to alkyd-based eggshell finishes

The 'antispatter' and flow requirements of such a product will almost certainly require use of one of the 'associative' type of thickeners now available. The film performance requirements are most likely to be met by harder acrylic polymer emulsions. Tinting with universal stainers will influence the choice of colloid, surfactants and coalescing and 'wet edge' solvents. Film properties may also be influenced by the type of extender which will affect the PVC required to achieve the required sheen level.

In this case it should be possible, from existing silk emulsion formulations, to arbitrarily select a grade and level of TiO_2 to give adequate opacity in two coats. For initial development a 'standard' extender used in silk finishes could be selected, leaving investigation of the influence of extender type on film properties to a much later stage. This will allow a detailed examination of the alternative resins identified, in conjunction with the alternative associative thickeners to establish which combination gives the most mar-resistant films, with good flow, levelling and antispatter properties when applied by roller. Simple colorant acceptance tests can be included in the evaluation of these paints to establish whether particular combinations are critical or preferred.

Having narrowed the choice of resin and thickener, the influence of extender on film properties can be reassessed and the influence of changes in surfactants, defoamers and solvents on colorant acceptance properties can be studied.

As with other more complex development projects, testing at the earlier stages should be confined to those properties directly influenced by the changes being made, in addition to the few basic tests such as viscosity, application and visual assessment of the dry film.

More complete testing against a full product test schedule should only be undertaken at critical stages in the development programme and particularly at the final stage of development, but before undertaking a works trial batch.

2.7 New classes of raw material

From time to time, the routine of normal development work is interrupted by the introduction of a new class of raw material. The temptation must be for the enquiring paint development chemist to drop everything and look at this new product for its potential. Before doing so, however, the possible benefits of the new material within the chemist's range of products must be questioned. The following questions should be asked:

1. Can it help to upgrade an existing product in such a way that a marketing advantage is gained? (Remember that it is also being offered to competitors.)
2. Can it help to achieve a cost reduction without loss of performance?
3. Can it help to develop a totally new product?
4. Would it be possible to handle in the factory without insurmountable problems?
5. Does it introduce a new handling hazard which would require special precautions?

Suppliers' representatives can be very persuasive (it is, after all, their job), but there is no point in undertaking any evaluation if the answers to the above questions are negative.

Assuming that some evaluation has been justified, much of the hard work will have been done by the raw material supplier before offering the new product to the market. Quite detailed formulating guidelines within representative type formulations and clear indications of possible interactions and preferred classes of other raw materials will normally be given. The main task for the paint formulator will be to evaluate the new raw material in the existing or proposed new formulations to confirm its potential. Within this, the formulator should identify the importance of the suggested interactions and preferred other raw materials. To what extent can the existing standard raw materials be used without sacrificing the benefits of the new raw material?

These factors can be established by direct comparison of experimental paints made up to the recommended formulation from the supplier, and equivalent paints incorporating existing standard raw materials.

Some other adjustments to the recommended formulation may be required to bring the product more into line with the company's own market expectations.

Guideline formulations issued for use in the North American market may be quite unsuitable for use in the UK or other European countries. However, this may be less critical in the case of a new product introduction, where expected 'norms' have not been established.

In many instances it may be sufficient to replace an existing raw material in an established product by the new raw material. Whilst this can be an acceptable short-cut in some circumstances, it may ignore other vital changes necessary to develop the full potential of the new raw material. In other words, negative results from such an evaluation do not necessarily imply that over-enthusiastic claims have been made for the new product.

It is also true that the more an established standard formulation has been developed and cost optimised, the more difficult it will be to show a cost benefit or performance benefit from a new, relatively unestablished raw material. Invariably, the recommended starting point formulations offered by the raw material supplier will be compared in the literature with fairly basic unoptimised standard formulations, which show the new product in a favourable light. Very often the smaller company with less product development resource can gain greater benefits from such new introductions than the larger company whose formulations are fully developed.

2.8 New technology

This is possibly the most interesting and challenging area of development for the paint chemist since it implies breaking new ground outside of those areas of formulation which may be regarded as 'established'.

While in many cases this will be linked to the development of new raw materials, opportunities exist for the establishment of new technology using available raw materials, from the paint and other industries, in an innovative way.

Examples of such innovative developments include such concepts as:

1. Paint-on central heating – uses include the potential for de-icing coatings for airport runways or fishing trawlers in arctic waters.
2. Multicolour coatings with decorative finishes ranging from subtle blends of related hues giving a textured appearance, through true multicolour coatings as used as antigraffiti coatings in public areas, to a 'brush-on' wood graining finish.

3. Self-stratifying coatings which layer during the drying process to give primer characteristics at the interface with the substrate, and top coat characteristics at the air interface.
4. Electrodeposition primers for use in the automotive industry to give more uniform protection of complex sections, particularly in normally inaccessible areas.

Similar opportunities still exist if the imagination to visualise the concept and identify a possible technological route, the commitment to the concept and the ability to sell it to commercial colleagues, and the willingness of a company to invest in the necessary development and launch costs, if successful, are forthcoming.

The challenge is there, and it can be a very stimulating, at times frustrating, experience to be involved in such developments. The main advice which can be given to any chemist who embarks on such a venture is

- Be committed.
- Be disciplined in approach.
- Plan well before acting at every stage.
- Maintain exacting standards and records.
- Set positive objectives, both for the chemist and team.
- Above all, maintain and convey enthusiasm.
- Know when to quit.

Bibliography

J. Boxall and J.A. Von Fraunhofer *Paint Formulation: Principles and Practice*, George Godwin Ltd., London (1980).

Oil and Colour Chemists Association (Australia) *Surface Coatings – A Complete Handbook of Paint Technology*, New South Wales University Press, Kensington, N.S.W., Australia (1974).

Oil and Colour Chemists Association (Australia) *Surface Coatings Vol 1 Raw materials and their usage; Vol 2 Paints and their applications*, Chapman and Hall, London Vol 1 (1983), Vol 2 (1984).

D.H. Parker *Principles of Surface Coating Technology*, John Wiley & Sons (1965).

H.F. Payne *Organic Coating Technology – Vol II. Pigments and Pigmented Coatings*, John Wiley, New York (1961).

G.P.A. Turner *Introduction to Paint Chemistry and Principles of Paint Technology*, Chapman and Hall, London (1980).

3 Raw materials

RICHARD WOODBRIDGE

3.1 Introduction

Essentially a paint consists of a *pigment* dispersed in a *resinous binder*, reduced to an acceptable application viscosity with *solvent*. One or more *additives* may be incorporated to modify one or other of the paint's film properties, application or storage characteristics.

Selection of the optimum class or grade of any one of the paint's constituents for use at the right level is the key to achieving a balanced product at the right cost. The word 'optimum' is used, rather than 'correct', since the choice will depend on many factors including local availability, relative costs, current usage, manufacturing facilities (e.g., design of the plant) and processes including bulk storage facilities and the ability to handle low flash materials. In practice the ultimate formulation achieved will be a compromise, with no two chemists arriving at exactly the same composition. It is this juggling act which provides stimulation and challenge to the paint formulator, but equally encourages a disciplined approach to ensure that the ultimate formulation and choice of raw materials is right for the company at that time.

Regardless of the type of raw material being selected, the question of local availability without cost penalty should be considered first. This particularly applies to materials used in bulk, since transport costs can be quite out of proportion to the cost of the actual raw material. This aspect is less critical with speciality additives, used at low levels, but even so should not be ignored. It has been argued previously that any formulation made should, as far as possible, be on the basis of raw materials currently maintained as a stock item; in other words, the addition of a new raw material to the stock inventory, or the need to review the opportunity for replacing an existing material with a new grade having enhanced properties should be justified.

Clearly these remarks do not apply when formulating an entirely new type of product, relative to the existing product range. In many cases, however, a new formulation is only a minor variation of an existing one and cannot justify the hidden costs of raw material diversification.

This discipline immediately reduces the number of potential grades of a given type of raw material to manageable proportions, and maintains the confidence of purchasing, manufacturing and financial colleagues. It is particularly important when the raw material is supplied and stored in bulk tanks – capital costs of providing additional bulk storage and handling systems may be quite unacceptable against the minor benefits of an alternative grade.

Having narrowed the field somewhat, there will still be a significant range of options for most raw materials. The aim of this chapter is to give some indications of the factors to consider in narrowing the choice still further. It is worth repeating that very few raw materials used in the paint industry can be regarded as commodity chemicals. Even when two materials have virtually identical nominal specifications, detailed evaluation may show a significant difference in a key paint property, depending on which grade is used.

3.2 Resins

3.2.1 *General*

Within the scope of this volume, it is only possible to give broad guidelines towards optimum selection. From there, selection of the preferred grade will be indicated by the detailed specifications of available grades from preferred suppliers and subsequent evaluation in a typical formulation.

The ways in which the vast range of resins may be classified leads to further complexity, as in the following:

- Low molecular weight v. high molecular weight.
- Solvent thinned v. aqueous v. powder.
- Air drying v. stoving v. radiation curing.
- Cross linking v. solvent evaporation.
- Solution v. emulsion or latex

Since this book deals more with practice than theory, it is appropriate to consider the choice of different resins according to their use in different market sectors, such as

- Decorative products: solvent thinned; aqueous coatings.
- Industrial finishes: general industrial; automotive etc.

Within the decorative market, it is essential that the resin can form a film under ambient conditions, i.e. is capable of drying within the temperature and humidity range normally experienced within the market area. Variations in ambient conditions in different areas can influence choice, for example emulsion or latex type resins relying on the evaporation of water for coalescence of the resin particles to take place may be quite inappropriate for use in an extremely cold or very high humidity environment. Similarly, with increasing concern on the potential health hazard associated with solvents (even aliphatic hydrocarbons), products designed for application to large areas of the interior of buildings are increasingly formulated on aqueous thinned latex resins.

Apart from these broad considerations, choice will be influenced by the nature of the product and its expectations, e.g. durability, gloss level, type of substrate, etc.

In the industrial paints market, the choice of resin will be influenced as much by the application and curing conditions available in the factory, as by its potential for achieving the required film performance. As industry moves towards more sophisticated finishing methods, such as electrodeposition of primers and ultraviolet (UV) or radiation curing of finishes, so will choice of the resin system become more critical.

Choice of type or grade of resin and level of usage will also depend on the required solids content of the coating being formulated. With increasing concern on the levels of emission of solvents into the atmosphere, the trend is towards high solids or solvent-free or aqueous coatings.

3.2.2 *Types of resins available*

Before considering selection of resins for specific types of end products, the general classes of resins available are identified and their broad characteristics considered.

3.2.2.1 *Alkyd resins.* Basically these are polyester resins produced by reaction of vegetable oil triglycerides, a polyol (such as glycerol or pentaerythritol) and a dibasic acid or anhydride (e.g. phthalic anhydride). Resins are usually categorised according to the oil content (oil length) and its nature. For example,

- Short oil (40%), e.g. dehydrated castor, coconut.
- Medium oil (40–60%), e.g. linseed, tall.
- Long oil (>60%), e.g. soya.

Other critical differences are in viscosity and solids content, acid value, hydroxyl value, the ratio of the three key ingredients and the type of solvent used for solution. Depending on the nature of the oil used and oil length, the resin may be capable of drying by autoxidation, or may require cross linking with products such as amino resins. Clearly, extreme diversity of products is possible and identification of suitable grades for a particular application must be made from the suppliers' detailed information.

Various so-called *modified alkyds* are also available, the basic alkyd being modified with urethane, polyamide, silicone, phenol formaldehyde or vinyl resins.

Urethane alkyds are prepared by replacing part of the dibasic acid by toluene diisocyanate. Long oil urethane alkyds, used in decorative paint formulations, give greater toughness and quicker drying than the unmodified alkyd.

Polyamide modified alkyds have thixotropic properties, produced as a result of hydrogen bonding. Available in a variety of grades with different gel strengths, they are used either alone or blended with unmodified alkyds to give improved application properties and sag control, mainly in decorative paints.

Silicone-modified alkyds are claimed to give outstanding gloss retention, particularly for exterior maintenance coatings for metal substrates, but their relatively high cost has limited their use to specialised applications.

Phenol formaldehyde modification gives increased hardness, and water and chemical resistance, but increases yellowing tendency, which tends to limit its use to primer formulations, particularly where the cost of epoxy resins would be prohibitive.

In vinyl-modified alkyds, the usual modifier is styrene or vinyl toluene, or, less commonly, methyl methacrylate. The prime benefits are faster drying, a harder film, and better colour, water and alkali resistance, offset by inferior gloss retention and solvent resistance. More recently, the high cost of vinyl toluene and supply difficulties on styrene led to the development of a range of fast dry alkyds for general industrial spraying and brushing enamels based on a modification using benzoic acid. Such resins are available in both aliphatic and aromatic solvents, the latter giving the fastest dry characteristics.

Special alkyds have also been developed in recent years, claiming much higher flexibility and retention of flexibility on ageing. These are particularly aimed at the exterior wood coatings sector, to counter growing concern about early failure of conventional alkyd coatings, particularly when used in situations where moisture can get into the timber.

3.2.2.2 *Polyester resins.* While the term polyester embraces alkyds as well as both unsaturated and saturated polyesters, it is normally reserved for oil-free acid or hydroxy functional polyester resins. Typically they are derived from di- or polyhydric alcohols reacted with di- or tri-basic acid or anhydride, dissolved in solvent. The balance between hardness and flexibility is controlled by careful selection of the prime constituents, together with modifications such as silicone. As it is much higher in cost than oil-modified alkyd, the use of polyester resins is restricted to high performance coatings such as coil coating, automotive base or clear coats and wood finishing. In stoving finishes they are generally cross linked by a melamine formaldehyde resin, while in air dry applications they are used as a two-pack formulation, cross linked with a polyfunctional isocyanate adduct.

Water-soluble polyesters and powder coating resins are also available.

3.2.2.3 *Acrylic polymers.* These are based essentially on acrylate and methacrylate esters, with possibly other unsaturated vinyl compounds; acrylic resins are available in both thermoplastic and thermosetting systems. In the latter case, additional functional groups are built into the polymer to allow cross linking with melamine formaldehyde resin (MF) or polyisocyanate. As a class, acrylic resins are noted for their excellent clarity, toughness, chemical and weather resistance, making them ideal candidates for automotive finishes, domestic appliance finishes and coil coatings for exterior cladding etc.

The solvents used depend on the composition of the acrylic polymer – shorter side chain polymers are relatively polar and require ketone, ester or ether–alcohol solvents. As the side chain lengthens, aromatic hydrocarbons may be introduced. The optimum solvent balance will also be influenced by the nature of the cross linking resin to be used.

Acrylic resins are, of course, also available in emulsion form, as latices or non-aqueous dispersions (NADs). These resins will be discussed separately.

3.2.2.4 *Amino resins.* Amino or nitrogen-containing resins are condensation products of compounds containing two or more amino groups, usually urea or melamine, with formaldehyde. While normally supplied as a solution in butanol/xylene, isobutanol or ethanol or mixtures of these, some grades are available in aqueous solution. Normally used as hardening or cross-linking resins with a range of hydroxyfunctional resins, they impart excellent colour and colour retention, hardness, and chemical resistance.

The lower cost urea formaldehyde (UF) resins are used in general purpose industrial finishes and acid-catalysed room temperature cured-wood finishes. The melamine formaldehyde (MF) resins, with their better chemical resistance, colour retention at elevated temperatures, better exterior durability and shorter baking schedules assist formulation of products with high flexibility, impact resistance, good gloss and excellent water resistance, particularly appliance enamels and automotive finishes.

3.2.2.5 *Phenol formaldehyde resins.* The most common PF resins used in coatings are based on the alkaline-catalysed condensation of a phenol and formaldehyde, known as a resole type resin. Various phenols may be used to give oil-soluble and oil-reactive resins, and resin-modified phenolics are used in varnish preparation. As general cross-linking agents, PF resins have been superseded by MF and UF resins on the grounds of infinitely superior colour. However, in applications where colour is not important, such as can coatings and tank and drum linings, PF resins are still used, either alone or in blends with alkyd or epoxy resins.

3.2.2.6 *Epoxy resins.* Normally these resins are based on the reaction between diphenylol propane (bisphenol A) and epichlorhydrin. Various grades are available with increasing molecular weight, melting point and epoxide equivalent. The structure of cured epoxy resins contributes to their high chemical resistance, toughness, thermal stability and rigidity. Their yellowing tendency limits their use to primer and undercoat compositions and can coatings where adhesion and chemical resistance are important.

While the low molecular weight liquid resins can be used as the sole resin, the higher molecular weight solid resins are used in conjunction with urea, melamine and phenolic resins. Two pack systems are also used, cross-linking with isocyanate, polyamines or polyamides. The latter systems give outstanding chemical resistance, hardness, abrasion resistance, flexibility and adhesion.

For pipeline, tank and marine coatings, epoxy resins are blended with reactive coal tar pitches, and cured with amine, amine adduct or polyamide resin.

3.2.2.7 *Isocyanates.* The isocyanate group of products is reactive at room temperature with groups containing active hydrogen atoms, such as alcohols, amines, phenols and water. Both aromatic and aliphatic polyisocyanates are available, the former being more reactive than the latter. Low molecular weight diisocyantes, such as toluene diisocyanate (TDI), are used for the preparation of urethane alkyds, whilst others may be used in the manufacture of blocked isocyanates used in electrocoat compositions, various solvent applied finishes and in powder coatings. Extreme care is necessary in the handling of isocyanates because the low molecular weight types are toxic and the vapours are irritants and sensitisers, even at extremely low concentrations.

Two-pack finishes based on hydroxy functional resins and isocyanate adducts are used as the basis for tough high solvent resistant coatings. Resins which may be used include alkyd, polyester, polyether, epoxy and acrylic, whilst a commonly used adduct is based on 3 moles of hexamethylene diisocyanate to 1 mole water. The two-pack alkyd finishes are used for wood finishing while the two-pack polyester finishes are used for high durability finishes for aircraft and the painting of plastic automotive components. Two-pack acrylic systems have been used extensively in automotive refinishing enamels.

3.2.2.8 *Silicone resins.* A range of alkyl and aryl organic silicone resins is available for modification of alkyd, polyester, acrylic and epoxy resins – the level of modification ranging typically from 15 to 40%. The major use of such silicone-modified resins is for the coating of steel coil for building cladding, due to the greatly enhanced durability.

Pure silicone surface coating resins are also available for applications requiring very high heat resistance properties.

3.2.2.9 *Vinyl resins.* These include homopolymers of vinyl chloride and copolymers of vinylchloride/vinyl acetate. The homopolymer polyvinyl chloride is normally used in the form of dried particles dispersed in a plasticiser (Plastisol) or a mixed solvent/diluent (Organosol). When applied and heated, the polyvinyl chloride particles swell and coalesce. The copolymer is soluble in ketones or blends of ketone and aromatic solvent. The resins are tough, elastic and water resistant with good extrusion properties – typical applications include the coating of strip metal for venetian blinds and bottle tops.

3.2.2.10 *Emulsion polymers/latices.* The term latex is strictly correct for these aqueous coatings, since they are dispersions in water of fine particle solid resins, such as 100% acrylics or copolymers of vinyl acetate (VA) with one or more of a whole range of alternative comonomers. These may include acrylates, vinyl versatate, ethylene, vinyl chloride or other vinyl esters. Copolymers of styrene with acrylics are also available. Particle size of such resins varies, dependent partially on method of manufacture and type of stabilisation, but typically would be between 0.1 and 1.0 μm, with most of the VA copolymers being around 0.5 μm. The finer particle latices, usually the surfactant stabilised 100% acrylics, tend to have inferior flow properties, but new technology in paint thickeners has largely overcome this weakness. Film formation is by coalescence of the latex particles, as the water evaporates, and in practical paints this process is assisted by the incorporation of coalescing solvents. Such resins are now used for a major percentage of interior and exterior finishes in the decorative paints market.

3.2.2.11 *Non-aqueous dispersions (NAD).* Analogous to emulsion polymers, fine particles (0.1–0.5 μm) of acrylic resin are dispersed in a non-aqueous continuous phase such as aliphatic hydrocarbon. A key to stability of such dispersions is the presence on the particle surfaces of an adsorbed layer of block or graft copolymer which, when solvated, acts as a barrier to flocculation – a process known as steric stabilisation. Addition of stronger solvents to the NAD can cause swelling of the polymer particle, leading ultimately to true solution. Such organic microgels are used in conjunction with thermosetting hydroxy acrylic polymers to improve rheological control during application. A major area of use of such NAD technology is in higher build automotive coatings, as sag control modifiers and assisting in orientation of metallic pigments.

3.2.2.12 *Water-soluble resins.* Certain resins, such as some formaldehyde condensates, including hexamethoxymethyl melamine and certain phenolics, are water soluble without further modification. Other resins can be made water soluble by building hydrophilic groups into their backbone – commonly acid or amine groups which are then solubilised by salt formation by the addition of volatile amine or acid. Maleinised fatty acids are commonly used as a modifier for epoxy or polybutadiene resins to achieve aqueous solution in the presence of an amine. Often a water miscible-coupling solvent, such as butyl cellosolve, is included to improve dilution characteristics. The presence of the amine tends to promote yellowing, while some concern exists about the toxicity of the coupling solvents, but such resins find application in metal primers. Alkyd and polyester resins formulated with high acid value can achieve water solubility. Water-soluble resins may be self-curing or may be cross-linked by the addition of soluble or emulsified cross-linking resins, such as water-soluble MF, UF or phenolic resins or beta-hydroxyalkylamides.

Epoxy resins may be modified by reacting with dicarboxylic anhydrides or trimellitic anhydride. Alternatively, they may be reacted with secondary amines, achieving solubility with acetic acid. Such resins would be cured with MF, phenolic or blocked isocyanate resins. The potential uses and development of new routes for achieving water solubility must accelerate with growing pressures for reduction of solvent usage.

A special class of water-soluble resins has been developed for use in primers applied by electrodeposition. Most of the foregoing types are suitable for adaption to electrodeposition, normally being used in stable dispersions of particles or micelles. Typical neutralising agents for anodic deposition are alkali or amines whilst lactic or acetic acid is used for cathodic systems.

3.2.3 *Selection of resins*

3.2.3.1 *Decorative products.* Decorative products covers decorative and protective products designed for use by either tradesmen or do-it-yourself (DIY) users, for on-site application to the exterior or interior of buildings. In this context buildings would include domestic properties, shops, offices, schools, etc., but would exclude the on-site application of coatings to industrial plant and structures, bridges, etc., which is normally classed as heavy duty maintenance painting. Whilst traditionally the decorative market was based on air drying oleoresinous media, during the last 40 years there has been a very significant move towards aqueous latex-based products, the actual percentage split varying somewhat from one country to another. The trend must be towards an ever increasing percentage of aqueous products.

Solvent-thinned products. Decorative gloss finishes are usually based on long oil air drying alkyds, dissolved in aliphatic hydrocarbons. Typical resins for quality finishes would be based on a linoleic rich oil with 60% oil length, using pentaerythritol as the esterifying alcohol. Viscosity grade would be determined by the required solids content in the finished product, taking into account the level of pigmentation required, according to colour and opacity requirements. It is common practice to incorporate a percentage of polyamide-modified alkyd resin, to introduce a degree of structure to the finished product, and fully gelled 'non-drip' formulations may be based entirely on such thixotropic resins.

Lower cost finishes may be formulated on medium oil length alkyds, but with the penalties of lower paint solids leading to lower film build and inferior covering power, inferior brushing characteristics and inferior weathering resistance. These disadvantages can be overcome substantially by the use of a medium/high oil length benzoic acid modified alkyd of higher viscosity.

Where a range of gloss finishes of varying quality is produced, it can simplify production to produce a common stabilised mill base (see Chapter 7), using a high quality lower viscosity alkyd as the dispersing medium, and to let down with various viscosity alkyds or thixotropic alkyd to achieve the different finishes required.

Primers and undercoats are based on similar concepts, but alternative oils which may produce some yellowing tendency are acceptable. For metal primers, shorter oil lengths are acceptable, giving faster dry and harder films, but these would not be recommended for primers or undercoats for exterior wood, where speciality high flexibility alkyds should be considered.

For products where quicker drying or greater toughness are required, urethane-modified alkyds should be considered.

Whilst in most countries exterior wall finishes are now based on latex technology, there are still areas where styrene or vinyl toluene (VT) modified soluble acrylic resins are used for masonry coatings. These also have the benefit of being resistant to early rain or high humidity during application.

Aqueous products. The choice of resin for a particular application will depend on the range available and relative prices in specific countries or markets. As a general rule VA copolymers are used for flat interior and exterior paints and general purpose silk finishes for interior use, while the more expensive 100% acrylics are preferred for high performance silk (aqueous eggshell) finishes, aqueous gloss and speciality primers and finishes for exterior wood, including so called 'house paints'. Styrene acrylate resins, often of a relatively fine particle size with high binding characteristics, are used for higher PVC flat coatings and for exterior masonry finishes.

Vinyl acetate/vinyl chloride/ethylene terpolymers, made by a pressure polymerisation process, are seen as alternatives to more conventional VA copolymers, but may gain ground in the formulation of (coalescing) solvent free low odour finishes for interiors.

3.2.3.2 *Industrial finishes.* Since the potential range of both resins and finished products is extremely wide, only very broad guidelines are possible within the scope of this volume. The choice of the preferred grade for a particular application must be influenced by local availability, cost expectations, methods of finishing and curing, expected performance and specifications, etc. Careful study of suppliers' product range and data sheets will indicate which specific grades are most likely to satisfy requirements and in most cases will give a good indication of formulating constraints. Since wide variations exist in the cost of different systems, it is important to be realistic and not to overformulate; success in this field is dependent on the right formulation at the right price. Listed below are the general recommendations found in the selection of resins for broad categories of products.

Heavy duty maintenance coatings. Air drying systems are required, normally using alkyd or modified alkyd technology, for durable exterior coatings. Urethane alkyds will give extra toughness and faster drying, while silicone alkyds will give maximum durability gloss coatings on metal. Phenolic modified alkyds may be chosen for increased water and chemical resistance, where colour retention is not important and costs of epoxy systems are prohibitive. Two-pack epoxy systems, cross-linked with isocyanate, polyamide or polyamines, would be the choice for high chemical resistance with extreme hardness, flexibility, adhesion or abrasion resistance. Where extremely high heat resistance properties are required, 100% silicone resins should be considered.

Appliance enamels. Basic requirements are good colour and colour retention with good build and resistance to alkalis and detergents. Recommendations include alkyd/amino systems, particularly using melamine formaldehyde as the cross-linker and possibly with inclusion of some epoxy resin. Other systems to consider are epoxy/polyamide, thermoset acrylic or acrylic/MF systems.

Exterior coil coatings. The main class of resin to consider is oil-free polyester, possibly modified with silicone for greater gloss retention. For stoving systems, MF would be used as the cross-linker. Alternatively, an isocyanate cured system may be used for air dry applications. Other possibilities are thermoset acrylic, fluoropolymers or polyvinylchloride plastisol systems.

General industrial paints (metal decorative finishes). The choice here is fairly wide, depending on required performance and application/curing conditions. For basic air drying systems, short oil alkyds or fast dry alkyds such as styrene, vinyl or benzoic acid modified resins would be the first choice. For higher performance, acrylic isocyanate or vinyl type resins may be used. Basic stoving systems would include alkyd/amino systems, with or without epoxy modification.

Machinery enamels. The main system used for fast dry and oil and petrol resistance would be fast dry alkyds or possibly a system based on chlorinated rubber.

3.3 Pigments

3.3.1 *General*

The main purpose of the pigment is to provide colour and opacity to the paint. Pigments will also contribute to film build, durability and corrosion resistance and assist in controlling other film properties such as level of sheen, toughness, mar resistance, permeability, intercoat adhesion etc., by adjusting the PVC of the film to the required level.

Pigment particles may range in size from 0.01–50 μm, or even larger in the case of granular fillers for textured masonry paints, but for the majority of applications a maximum diameter of 25 μm would be used. The pigments may be supplied as dry powder or in the form of pastes or slurries, but regardless of type the main objective within the paint formulation and manufacturing process is to ensure that the chosen pigments are fully dispersed, wetted and stabilised, to prevent flocculation in the can or during the drying process. This aspect is covered fully in Chapter 7.

The various pigments available are normally considered within the following groups:

- White pigments.
- Coloured pigments.
- Extenders.
- Anticorrosive pigments.

3.3.2 *White pigments* (titanium dioxide)

Whilst in the past basic carbonates of lead, zinc oxide, antimony oxide, and lithopone (zinc sulphide/barium sulphate) may have been considered as important members of this class, today almost without exception only one white pigment is used, that is titanium dioxide. However, as has been stated previously, rarely can basic paint raw materials be considered as commodities and within available supplies of titanium dioxide many grades with quite dissimilar properties are found. Selection of the optimum grade for any application is important, particularly in view of its cost contribution to the finished product. Whilst two main manufacturing routes exist, i.e. sulphate and chloride processes, this is less important than the post treatment received, which conditions the surface of the pigment to confer particular properties for specific applications.

It should also be noted that the pigment is available in two crystalline forms, rutile and anatase. However, in normal paint applications only the rutile grade is used, being much more durable and having higher refractive index, higher specific gravity and greater chemical stability. As measured by tinting strength, rutile pigment has a potential hiding power approximately 45% greater than anatase.

The surface coating used on the titanium dioxide may range from 0–20% of the pigment, as supplied. The major portion of the coating is inorganic and may consist of Al_2O_3 , SiO_2 , or ZrO_2 or a combination of these. Additionally, some organic treatment may be applied to improve dispersibility in a variety of media, typically polyols, amines and silicone derivatives. Ultimate pigment characteristics may be influenced also by the way in which the coating is applied.

The nature and weight of the coating will influence particle size (around 0.2 μm) specific gravity, bulk density and particularly surface area, which affects oil absorption or water demand. Other important differences will be in colour (undertone), durability and dispersability. Suppliers give their recommendations on which grades should be considered for particular type coatings, including such categories as:

- Superdurable grades for coil coatings and automotive finishes.
- Highly durable grades for exterior gloss finishes for decorative or industrial applications.
- Grades with excellent optical properties for use in appliance enamels, water-based industrial paints, etc.
- Highly coated grades (80–86% TiO_2) for maximum opacity in flat emulsion paints.
- Multi-purpose grades to give optimum properties over the widest possible range of applications.

In practice, the distinction between some grades is very small and unless it is necessary to highlight a particular feature, as in the first two categories above, the choice of grade may be very wide. Increasingly, with bulk handling and storage of titanium dioxide, there are benefits in identifying a suitable multi-purpose grade which may be used in a wide range of products. In the manufacture of a decorative product range, it is increasingly recognised that suitable multi-purpose pigments may be used in both solvent thinned exterior gloss finishes and aqueous latex coatings ranging from low to high PVC. The argument that highly coated grades are preferred for medium or high PVC matt emulsion paints has been challenged and enamel grades are used increasingly in such paints, particularly in conjunction with opaque polymer.

Apart from the selection of the optimum grade, the other critical decision is to establish the optimum level of use. This is clearly determined by the required level of opacity, but will be influenced by other factors such as the PVC, the degree of wetting and dispersion achieved and maintained, the applied film thickness and paint solids content, and presence of tinting or other functional pigments.

In conventional white air drying gloss finishes, the TiO_2 level may vary from 350 kg/1000 l for premium quality finishes to 240 kg/1000 l for economy products, maintaining a PVC of approximately 17.5% or p/b ratio of 0.75:1, within coatings having total solids contents ranging from 70–50%.

In decorative latex finishes, below the critical pigment volume concentration (CPVC), the TiO_2 level may vary from 200–300 kg/1000 l depending on the level of gloss/sheen, presence of other opacifiers, such as opaque polymer or calcined clay, overall paint volume solids, etc. Above the CPVC, TiO_2 levels are progressively reduced as dry hiding increases, and may fall as low as 20 kg/1000 l in some economy products.

In high build textured exterior coatings, TiO_2 levels should be reduced in proportion to increased film build, i.e. a standard level of TiO_2 usage per unit area of surface covered should be maintained.

3.3.3 *Coloured pigments*

The traditional first classification of coloured pigments is into inorganic and organic types. In broad terms, the basic properties of each class are summarised in Table 3.1 (Lambourne, 1987).

Table 3.1 General properties of inorganic and organic pigments

Pigment property	Inorganic	Organic
Colour	Sometimes dull	Usually bright
Opacity	Normally high	Relatively low
Colour strength	Usually low	Normally high
Fastness to solvents – 'bleed resistance'	Good	Varies widely from good to bad
Resistance to chemicals	Varies	Varies
Heat resistance	Mostly good	Varies
Durability	Usually good	Varies
Price	Relatively inexpensive	Varies, but some are very expensive

(Ref. Lambourne: *Paints and Surface Coatings and Theory and Practice* p.113)

It has also been shown (Lambourne, 1987) that over 5000 different coloured pigments, representing over 700 different chemical types, are available and listed in the *Colour Index* published by the Society of Dyers and Colourists in association with the American Association of Textile Chemists and Colorists.

Within this brief introduction it is not possible to give specific recommendations for the selection of pigments for products of different types and colours. In general, the choice of coloured pigments is determined by the ultimate colour required, but also takes into account requirements for opacity, durability, heat and solvent resistance, toxicity requirements and cost. The summary of broad properties listed in Table 3.1 indicates that there is a strong case for using inorganic pigments wherever possible, but this may not be feasible where very clean tints or bright deep shades are required.

Care must also be taken when selecting blends of pigments of widely differing size, surface treatment or specific gravity, where incorrect choice may lead to problems of flooding or flotation in the applied film during drying. Selection of pigments with an appropriate undertone is important in achieving the required shade, particularly if it is very clean. The listing of main properties and representations of full shade and reductions with white, published by the pigment manufacturers, may act as a guide.

Consideration must also be given whether to use dry pigments, dispersing in the chosen media, or predispersed pastes, available in more restricted colour ranges in solvent, oil, aqueous or universal media. In a situation where specific ranges of pigments are regarded as the standard range stocked within a company it is possible to set up basic data within the storage bank of a computerised match prediction program. Such programs can give alternative options at a standard level of opacity in increasing order of cost, together with an indication of the degree of metamerism likely. In a very competitive market, use of such techniques is strongly recommended since the savings achieved as a result of optimal use of coloured pigments can justify the costs of establishing the system.

3.3.4 *Extenders*

While, historically, extenders were used as a means of adding low cost bulk to a paint's solids content, hence the popularity of barytes when paints were sold by weight, today the range of extenders available is extremely comprehensive and correct choice can influence many of the paints' properties, including gloss, opacity, flow and levelling, durability, film toughness, permeability and rheology.

Chemical composition is far less important in determining the influence of extenders on paint properties than physical characteristics, such as particle size and shape, oil absorption, specific gravity and the nature of any surface treatments or calcining.

Particle size may vary from less than 0.1 μm to over 3000 μm, but the majority of commonly used extenders would be in the 0.2–20 μm range. Very fine extenders have high surface area and hence very high oil absorption, while the relatively coarse extenders (> 20 μm) have an adverse effect on film smoothness and lower opacity due to the 'window effect'. Coarse extenders have a place in textured finishes, non-slip coatings, etc. Particle shapes may also vary widely from nodular crystalline forms to amorphous aggregates, needle or rod-like particles and lamellar plate-like particles. Calcined extenders may have highly fissured surfaces which remain unwetted by binder within the film, creating light scattering sites which contribute to dry hiding. Careful selection of extender blends can influence film packing, and give positive benefits in film strength, with consequent improvements in wash resistance and durability.

Apart from particle size and shape, the other most important characteristic in correct selection is oil absorption. The higher the oil absorption, the lower will be the coating's critical pigment volume concentration (CPVC.) Since, on a volume basis, extenders generally are much cheaper than TiO_2 or binder, cost effective formulations require as high CPVC as possible. Also, for coatings sold by volume, there are advantages in using extenders with as low specific gravity as possible. These considerations explain the popularity of medium particle size whiting (calcium carbonate) which is relatively low cost and has low specific gravity and oil absorption. However, to formulate entirely on whiting as the extender may be acceptable for low cost interior coatings, but requires blending with others such as china clay, talc or mica to improve toughness and durability in higher performance coatings.

The most commonly used extenders are listed in Table 3.2, together with basic data on refractive index (RI) and specific gravity (sp.gr.) and an indication of the normal limits of oil absorption and size range, depending on grade. It will be noted that barytes or blanc fixe, whilst having much higher specific gravity than all other extenders, also have the lowest oil absorption. This makes them a logical first choice of extender for inclusion in alkyd-based gloss, undercoat or primer formulations to increase the volume solids with minimum effect on gloss or film porosity.

Table 3.2 Main types of paint extenders

Common names	Chemical type	Specific gravity	Refractive index	Oil absorption* (g oil/100g pigment)	Particle size† ranges (μm)
Barytes, blanc fixe	Barium sulphate	4.4	1.64	10–12.5	10–30
Whiting, chalk, paris white, calcite	Calcium carbonate	2.7	1.5–1.6	10–30	0.5–30
Dolomite	Calcium magnesium carbonate	2.85	1.62	15–19	0.5–30
Wollastonite	Calcium silicate	2.9	1.63	19–21	1–30
Minex, syenex, nephaline syenite	Anhydrous sodium potassium silicate	2.61	1.53	21–29	2–20
China clay, kaolin	Aluminium silicate	2.6	1.56	33–48	0.2–25
Talc, asbestine	Magnesium silicate	2.75	1.56	35–65	0.5–20
Calcined clay	Calcined aluminium silicate	2.5	1.56	50–80	0.5–15
		2.63	1.62		
Mica	Alumino silicate	2.84	1.58	50–75	1–50
Dicalite, celite, diatomite	Silica	2.33	1.46	120–210	1–50

*Individual paint grades have a narrower specification, within these ranges.
†The lower oil absorption figures quoted apply to the larger particle size grades.

Plate-like extenders such as china clay or mica can have a reinforcing effect on the film, giving improved washability in flat interior coatings and improved durability in exterior coatings. China clay, particularly the calcined grades, contributes to opacity and can also assist in giving a soft structure to the paint, reducing the tendency to drip during application. Talc is also good in this respect and gives greater burnish resistance than china clay. A combination of whiting, clay and talc is generally used as a mixed extender system in quality flat emulsion paints, whilst micro-dolomite or minex may be used as part of the system in smooth exterior coatings. For fine-textured exterior coatings, a mixture of whiting or micro-dolomite with quartz and mica can give an extremely durable system. There are no hard and fast rules on the selection of extender blends; much depends on local availability, relative prices, colour requirements and the extent to which the extenders, rather than other additives, are used to control the rheology of the system. However, the right blend for a particular system can influence the required balance of properties in both the liquid paint and the dry film. Extenders are no longer regarded as low cost additives but as vital building blocks to a successful paint formulation.

The high oil absorption extenders such as fine particle silicas can be used as flatting aids, since only low levels are required to reduce the level of sheen significantly. Medium to high levels of such extenders should be avoided since they have a very significant effect in lowering the CPVC of the system and also can adversely affect flow.

As will be demonstrated in Chapter 4, one particularly valuable class of extender in achieving cost-effective formulations is the so-called spacing extender. To be effective these require high brightness, have relatively low oil absorption (due to a nodular shape) and have a mean particle diameter of approximately 0.8 μm. Certain grades of calcium carbonate meet this requirement.

3.3.5 *Corrosion-inhibiting pigments*

Traditionally, the most effective anticorrosive pigments have been based on lead or chromates, both of which are now causing concern on the grounds of toxicity. This is leading to the growing use of other types of pigments or to the replacement of traditional primers by other technologies, such as electrogalvanising. The rate of change to alternative materials varies from one world market to another, leading to significant differences in approach between countries.

Zinc dust is now the most widely used anticorrosive pigment worldwide, particularly in 'zinc-rich' coatings (where pigment content is in the order of 95% of the total weight of dried film).

Redlead still retains high usage, in markets other than the USA, particularly in heavy duty maintenance coatings for structural steel, with end-uses such as petrochemical installations, power generators, civil engineering, etc.

Calcium plumbate is still regarded as the most effective pigment in primers on zinc or galvanised iron surfaces, due to the formation of soaps which give high adhesion and toughness to the film.

Zinc and strontium chromates have been used, often in conjunction with red iron oxide, for both decorative and industrial general purpose metal primers. A more specialised application for zinc tetroxy chromate is in 'wash-primers' for aluminium surfaces, particularly on aircraft, where alternative materials have not yet been adequately proven.

Basic lead silico-chromate has been widely used in coatings for both the structural steel industry and for automotive paints.

Calcium, strontium and zinc molybdates are currently regarded as non-toxic, and are gaining in interest as potential replacements for chromates.

Zinc and calcium phosphate are also regarded as non-toxic and, being white, allow the production of white or pale colour primers. They may be used in aqueous and solvent-thinned systems and are increasingly employed as a result of proven performance.

Barium metaborate is also suitable for use in both aqueous and solvent-thinned primers, and with proven good performance, but its growth is restricted somewhat by concerns on level of toxicity.

Lamellar pigments are used to increase the barrier properties of the film as it is recognised that this can inhibit corrosion by reducing the passage of oxygen and moisture to the substrate. Micaceous iron oxide is growing in importance as an alternative to red lead in paints for structural steel and particularly for bridges. Other lamellar pigments used to enhance the properties of anticorrosive pigments are aluminium flake, mica and, more recently, glass flakes.

3.4 Solvents, thinners and diluents

3.4.1 *General*

The main purpose of the solvent is to dissolve the resin, whilst thinners and diluents reduce the viscosity of the paint to an acceptable level for application. Choice of the optimum solvent or solvent blend will also influence the flow of the coating and the rate of drying, at least in the early stages.

While the above describes the role of solvents in paints based on solution resins (solvent-thinned paints), consideration must also be given to the case of aqueous products, where the resin is dispersed and the water acts as a diluent, while 'coalescing solvents' assist in the fusion of latex particles as the water evaporates. Similarly, in NADs, aliphatic hydrocarbons act as a diluent for the dispersed resin.

Factors influencing choice of solvent for any specific product are solvency, boiling point or range, evaporation rate, flash point, odour, toxicity and cost. In practice, it is found that a mixture of solvents/diluents is necessary to give the optimum results in any application. Various techniques have been developed to help establish optimum blends with the required solvency, known as 'solubility parameter systems'. For a detailed review of such an approach reference should be made to Lambourne (1987).

The main types of solvents used and typical examples of each class are shown in Table 3.3.

3.4.2 *Solvents in air dry finishes*

Conventionally, long oil alkyds are dissolved in aliphatic hydrocarbons such as white spirit, with possibly 1–2% of dipentine to assist in anti-skinning. Products used on large interior surfaces such as alkyd-based eggshell finishes may use low odour aliphatic solvents, and, increasingly, solvents with very low aromatic content to reduce potential hazards. Shorter oil length alkyds, used in metal primers or finishes, may use a solvent with higher aromatic content to improve solubility and give faster dry. Certain fast dry alkyds may be thinned entirely with an aromatic solvent such as xylene.

Table 3.3 Main classes of solvents

Aliphatic hydrocarbon	White spirit, distillate
Aromatic hydrocarbon	Xylene, toluene
Alcohols	Methanol, ethanol, propanol, butanol
Ketones	Acetone, methylethyl ketone, methyl isobutyl ketone, diacetone alcohol
Esters	Ethylacetate, butyl acetate
Glycol ethers	Butyl cellosolve, butyl carbitol, cellosolve, propasol
Glycol ether esters	Butyl cellosolve acetate, butyl carbitol acetate

3.4.3 *Solvents in cross-linked and stoving finishes*

Choice of the optimum blend will be influenced by the nature of the resin system, the application method and the curing conditions. The number of permutations is so vast that it is impossible to give any generalised recommendations. Clearly the system used must have adequate solvency for the resin and maintain the resin in solution during the 'flash off' stage. Guidance towards acceptable systems should be given by both resin and solvent suppliers. In systems cross-linked with polyisocyanates it is important to avoid solvents containing OH groups, or contaminated with water, since these could react with the NCO groups.

3.4.4 *Solvents in aqueous finishes*

In latex-based finishes the main requirement for solvent is to assist the coalescence of the dispersed resin particles after evaporation of water. While glycol ethers and glycol ether esters tended to be used in the past, Texanol® is now widely favoured because it is a powerful coalescent for both acrylic and vinyl copolymer resins, and is insoluble in water. Approximately 10% Texanol is commonly used on resin solids. White spirit is often used as the coalescent in finishes based on styrene–acrylate resins.

In addition to the coalescing solvent, glycols such as ethylene glycol or propylene glycol are used at a level of 2–5% on total paint to assist in control of wet edge time and flow out. In the case of aqueous gloss paints it is recommended that the pigment should be dispersed in the glycol solvent rather than in water, to assist in achievement of minimum flocculation and hence maximum gloss.

3.5 Rheology modifiers

3.5.1 *General*

Control of the rheology of a paint is essential if a product with good appearance in the can, free from pigment settlement, with good application properties, good flow and levelling and freedom from defects such as sagging is to be achieved.

A basic understanding of rheology is therefore necessary and in particular of the relationship between shear rate and these various coatings' properties.

Viscosity, or resistance to flow, is defined as the ratio of shear stress to shear rate, i.e.

$$\text{Viscosity} = \frac{\text{Shear stress}}{\text{Shear rate}}$$

Shear stress is the force applied to cause flow, whilst *shear rate* is the rate at which one layer of liquid flows past adjacent layers. For the purposes of this study the relationships between viscosity and shear rate are considered. These can be illustrated simply in a viscosity profile diagram (Figure 3.1). From this various types of flow are recognisable.

- *Newtonian flow*, where viscosity is independent of shear rate – typical of oils and solvents.
- *Dilatancy*, where viscosity increases with increasing shear rate – typical of highly pigmented systems.
- *Pseudoplastic flow* where viscosity decreases as shear rate rises – typical of many latex coatings.
- *Thixotropic flow*, similar to pseudoplastic flow but in which there is a time delay before the viscosity recovers, on reducing shear rate or ceasing shear. The area between the two curves, of increasing and then reducing the shear rate, is a measure of the 'structure' of the system.

The viscosity profile in Figure 3.2 shows the relationship between shear rate and important paint properties. It will be noted that for the control of application properties, by brushing, spraying or rolling, viscosity at shear rates of $10^4\ s^{-1}$ are important. On the other hand, pigment settlement is influenced by viscosity at very low shear rates of between 0.001 and $0.1\ s^{-1}$. Flow and sagging is influenced more by viscosity at shear rates of between 0.01 and $1.0\ s^{-1}$, while appearance in the can and brush loading is more a function of medium shear rates.

It follows that the ideal paint should have a degree of thixotropy which gives high viscosity at rest, preventing settlement, but in which the shearing forces during application create high shear rates, reducing the paint's viscosity to a low level (50–250 mPa s (centipoise), depending on the type of paint) to give easy spreading over the surface. A delay in recovery of structure allows good

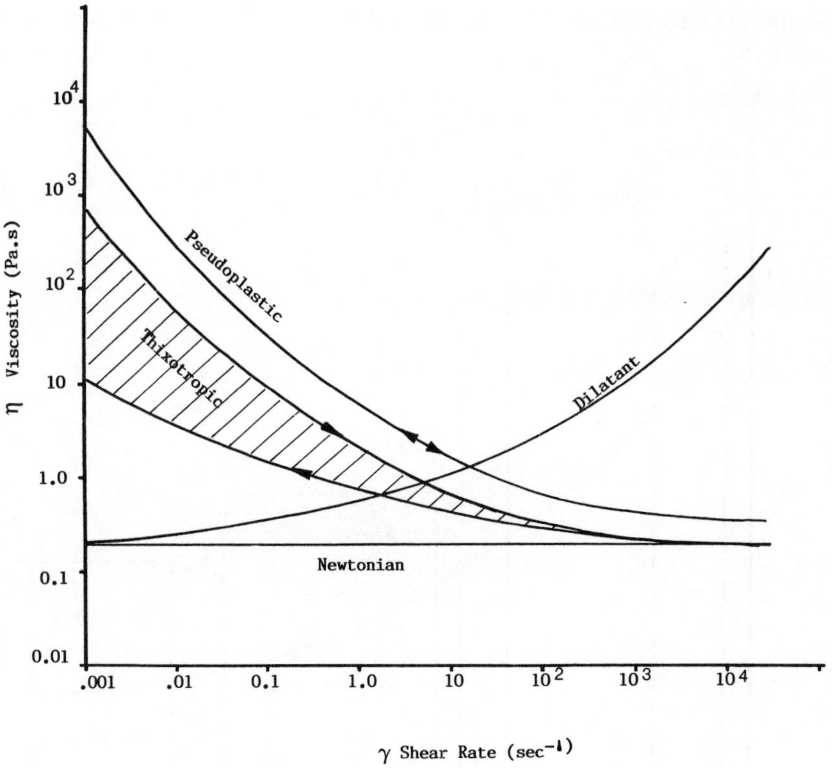

Figure 3.1 Viscosity profiles.

levelling and flow out of brush or roller marks, but should be sufficiently rapid to prevent sagging.

In products based on solution resins, viscosity tends to be more Newtonian, with the actual level being influenced by resin solids content and the resin viscosity (controlled by molecular weight). Higher levels of pigmentation will influence this viscosity and introduce more pseudoplasticity. In latex paints, the resin is suspended as fine particles and despite its high molecular weight does not influence the viscosity of the continuous phase (water). This is affected more by the type of colloid used for stabilisation and level of pigmentation, both of which have more influence on low shear rate viscosity. High shear rate viscosities of such products tend to be quite low, in the range 50–70 mPa s.

In both solvent- and water-thinned products there is, therefore, a need for modification with specialist materials to assist in achieving the ideal viscosity profile for individual products. Such materials are generally classified as rheology modifiers. Since the various technologies available tend to be specific to either solvent- or water- thinned products, they will be considered separately.

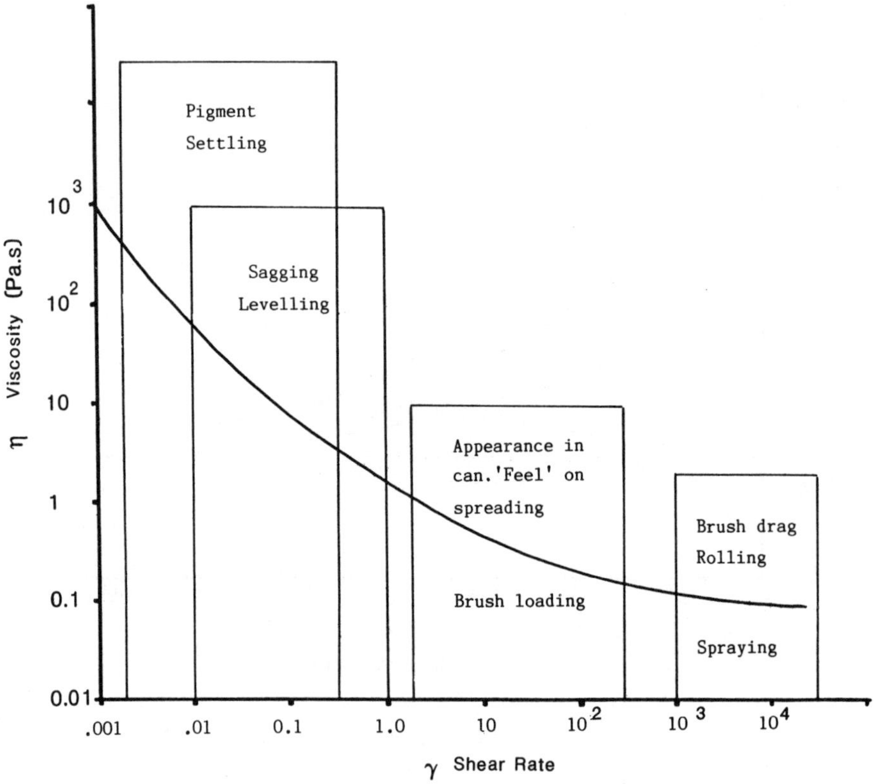

Figure 3.2 Relationship between shear rate and main properties.

3.5.2 *Solvent-thinned products*

The polyamide modified resins are used at the macro level of modification, normally in conjunction with 'liquid' resins to achieve a desired level of structure. Commonly 10–20% of the total resin system will be replaced by so-called 'thixotropic alkyds', to give a pleasing soft structure to the appearance of the product with good control of settlement and sagging. For the fully gelled 'non-drip' product for the DIY user, up to 100% of the resin system may be of the polyamide modified type. It must be recognised that introduction of such resins tends to lower gloss and durability, in addition to reducing overall paint solids and film build, hence the use of lower levels in premium products for the trade decorator.

The alternative route to increasing low shear rate viscosity is by the incorporation of relatively low levels of additives, which tend to fall into two classes:

1. Organoclays, such as bentone.
2. Organic modifiers.

For maximum efficiency of the bentone clays and full gelation, the individual platelets of the clay must be fully dispersed and solvated – normally achieved by a polar activator such as low molecular weight alcohols, in the presence of a small amount of water. The bentone may be used as a pre-gel or dispersed *in situ*, depending on the pigment dispersion process being used. Often the bentone is used in conjunction with a surfactant such as soya lecithin, typical levels being 0.15% of each on total paint weight.

A wide range of organic thixotropic modifiers is available, in powder, paste or liquid forms, the former grades generally requiring shear, heat and time to achieve maximum gelling efficiency. Choice of grade depends on the type of resin system being used and particularly on whether the polarity of the media is low, medium or high. Levels of use also vary widely from 0.3–4.0% of total weight, depending on the grade used and the nature of the finished product. Detailed guidance must be obtained from the manufacturer's literature. To achieve optimum control of all properties influenced by viscosity at low shear rates, a combination of two thixotropes is often used.

3.5.3 *Water-thinned products*

Within this area a very large number of alternative materials is found, ranging from the various grades of protective colloids used to impart stability to the formulation, to the true rheology modifiers such as the associative thickeners, and to the water-dispersible montmorillonite clays such as bentonite and laponite. Most of these materials are water soluble and choice of the optimum grade and level of use is important in achieving the right balance of properties and maintaining stability on storage. It is also important to recognise that different materials will give very different rates of recovery of structure after shearing, influencing flow-out of brush marks or roller pattern. Other significant differences between grades are in their effect on roller 'spatter' and in their susceptibility to biodegradation which may influence choice of type and level of bactericide required. Since these materials are such an important part of latex paint formulation, they are now considered in more depth.

3.5.3.1 *Cellulosic thickeners.* Within this class of material, the formulator is confronted with a bewildering choice of grades. Chemical types may range from the anionic sodium carboxymethyl cellulose (SCMC) to various non-ionic types such as methyl cellulose (MC), methylhydroxyethyl cellulose (MHEC), methyl hydroxypropyl cellulose (MHPC), hydroxyethyl cellulose (HEC) etc. Most types are available in different viscosity grades ranging from very low (10 cp for a 2% solution at 25°C – measured at low shear rates by Brookfield) to extremely high (100 000 cp). Grades are also available with varying degrees of substitution, which designates the average number of hydroxyl units on the anhydroglucose unit which have been reacted with ethylene oxide. A further variable is the physical form, which may be powder or granular, or an easy dissolving grade where solution characteristics have been retarded by the addition of a small amount of glyoxal.

In considering these various options availability and cost must be taken into account, while being aware of the key differences in their characteristics. In practice, the most commonly used types in paint are medium viscosity grades of SCMC and HEC, both of which remain as stable solutions over the temperature range 0–100°C, by comparison with some of the other non-ionic grades which precipitate or gel above their critical temperature (from 50°C upwards). HEC has a greater tolerance to pH changes than SCMC, which is most stable over the range 7–9. Typical paint grades of SCMC have relatively low degrees of substitution (0.7–0.9), whilst the HEC grades are fairly highly substituted (2.5). The higher degree of substitution gives greater resistance to biodegradability, but nevertheless it is normally advisable to add an effective preservative at the time of making a solution, and especially if the solution is to be stored or pumped through a dedicated pipeline system. Failure to adequately preserve can lead to the production by microbial attack of cellulases, which can rapidly degrade the viscosity of the solution or paints based on it.

While medium viscosity grades are preferred for higher quality paints, there is a tendency to use high viscosity grades in low-cost high PVC paints. The higher viscosity grades tend to be more thixotropic than lower viscosity grades, leading to less flow and to a less creamy structure in the can. This is offset by their reduced effect on water sensitivity due to the lower content required for a given viscosity. Typical levels used range from 0.25–0.6% on total paint, depending on the viscosity grade used, and the nature of the formulation.

The addition of colloid can be made as a dry powder preceding the addition of extenders and pigments at the dispersion stage, or as a pre-made solution (normally 2–2.5%), on completion of the dispersion stage but before addition of emulsion or latex. Use of the delayed solubility or granular grades can improve the quality and speed of solution, avoiding the formation of large gel-like solvated flocculates. While incorporation at the dispersion stage eliminates the need for separate solution making and storage facilities, the presence of the colloid impairs the achievement of optimum wetting and dispersion of the pigment and the solution route is strongly recommended if at all possible.

Mention must also be made of the more recently introduced hydrophobically modified HEC, in which long chain alkyl groups are chemically bound to the polymer backbone. While traditional cellulosic colloids build viscosity as a result of hydrogen bonding, the hydrophobic modified HEC additionally builds viscosity by association of the hydrophobic groups – hence the term 'associative thickener'. Paints thickened with such colloids are less pseudoplastic than paints thickened with traditional colloids, giving lower viscosity at low shear rates (and hence improved flow) but higher viscosity at high shear rates, resulting in improved film build when applied by roller or brush (and hence better hiding). A further particular benefit of this colloid is its high resistance to spatter when applied by roller. Although this material is relatively new, initial indications are that it offers a number of formulating benefits and improved paint stability over other synthetic associative thickeners, particularly when used in vinylacetate copolymer systems. In surfactant stabilised acrylic systems the urethane type thickeners are more appropriate.

3.5.3.2 *Acrylic thickeners.* Based on alkali soluble acrylic polymers, these thickeners are usually supplied in emulsion form at a low pH, which allows a much higher solids content than would be possible if supplied as a solution. Normally added to the paint as a post addition, the thickener emulsion is usually pre-diluted to approximately 15% solids content and added slowly with constant stirring to the paint which should be pre-adjusted to pH 8.5. Typical levels of 'solid' thickener resin range from 0.5–2.5% on total paint.

Acrylic thickeners tend to give much more Newtonian viscosity than traditional cellulosic thickeners, with good flow characteristics. However, this can be influenced by the particular binder used and choice of coalescent solvent. Texanol has been found to produce the best flow performance. Particular benefits of the acrylic thickeners are their minimal effect on gloss of a latex coating and their high resistance to biodegradation. The combination of good flow and gloss characteristics makes them a natural choice in latex gloss systems. Certain grades also give extremely good spatter resistance.

On the negative side, since the products are alkali soluble, they can have an adverse influence on alkali resistance, recoatability and wet adhesion of the finished coating.

3.5.3.3 *Urethane type associative thickeners.* The development of this type of non-ionic associative thickener allowed for the first time the attainment of acrylic latex coatings with:

- alkyd like flow and levelling, due to lower low-shear viscosity;
- increased film build, due to higher high-shear viscosity;
- excellent spatter resistance.

The side chains present form an 'associative' network with themselves and with the binder particles, and to a lesser extent with the pigment. This network has the effect of increasing high shear viscosity while not showing the same influence on low shear viscosity as is caused by conventional cellulosic colloids.

Normally urethane thickeners are used in combination with thickeners having a greater influence on medium and low shear viscosity to give controlled viscosity over the whole range. Intermediate viscosity may also be influenced by the presence of 'active' constituents in the paint, such as water-miscible solvents or surfactants. Of the various water-miscible solvents, butyl carbitol has the greatest effect on reducing medium shear rate (stormer) viscosity. Water-immiscible solvents such as texanol have virtually no effect. In the case of surfactants, the viscosity reduction effect is greatest with anionic surfactants and higher HLB value non-ionic surfactants. Glassy phosphate type sequestering agents such as Calgon® or Tetron® should be avoided since they almost totally destroy the thickening efficiency of urethane associative thickeners.

In establishing the correct balance between urethane and co-thickeners, it should be ensured first that the coalescing and other solvents and surfactants have been optimised; the level of urethane thickener to give the required high shear viscosity is then established. Additions of alternative less efficient urethane or cellulosic colloids such as HEC can be made to increase the medium shear rate viscosity, or butyl carbitol added to reduce it. Care must be taken not to add sufficient cellulosic colloid to upset the flow characteristics at low shear rates.

The various suppliers of these thickeners give good guidelines on their optimum use, but it must be recognised that the interactions and hence viscosity profile will be influenced by the nature and grade of latex used, by the non-volatile content of the formulation and by the presence of other 'active' materials, such as tinting pastes. Only by experimentation within typical formulations will the preferred route to optimisation of the whole viscosity profile be found.

3.5.3.4 *Montmorillonite clays.* Materials such as bentonite can be used at low levels (typically 0.1–1.0%) to enhance low shear viscosity and hence sag control. Care must be taken to avoid too high a level which would introduce very poor flow, although this effect is influenced also by the nature of the main thickener system, e.g. use with HEC will give higher low-shear viscosity than use with SCMC.

3.5.3.5 *Organo titanates.* To achieve higher levels of structure or even gel characteristics, materials such as triethanolamine titanate (TET) may be used at levels of 0.1–0.5% on total formulation. The actual level will depend on the nature of the prime cellulosic colloid and the influence on structure of the surfactant system chosen.

3.6 Additives

3.6.1 *General*

By implication, additives are materials added, normally in relatively small quantities, to control one of the paint's properties. A number of suppliers now offer a comprehensive range of materials to control particular paint defects, and it must be a natural temptation to reach for the appropriate bottle on the shelf when faced with a specific problem. It should always be recognised, however, that a good formulator will attempt to eradicate the problem by adjustment of the basic formulation rather than resort to 'corrective' additives. Such materials should only be used as a last resort and never as a matter of routine.

Not all additives fall into the category of 'corrective' – many form an essential part of a well balanced formulation but are included in the category of additives because they are used in small quantities, relative to the main ingredients.

The various broad classes of materials used to control properties as either positive (constructive) or negative (corrective) additives are shown in Table 3.4

Table 3.4 Paint additives

Constructive	Corrective
Dispersing/wetting agents	Anti-skin
Driers	Anti-flood/float
Preservative/biocides	Anti-foam
Insecticides	Anti-gassing
Can corrosion inhibitors	Anti-static
Flash-rust inhibitors	
UV absorbers	Anti-freeze
Reodorants	

3.6.2 *Dispersing/wetting agents*

As their name implies, these materials are added at the dispersion stage of manufacture to assist in the wetting/dispersion of the pigment. The range of materials available is very wide and choice depends on the type of pigments and media being used. Known also as surfactants (surface-active agents) they vary from anionic to non-ionic, cationic or amphoteric. As a general rule anionic types may be preferred for use with inorganic pigments while non-ionic types tend to be selected for the dispersion of organic pigments. Whichever material is selected as optimum, it should be ensured that only the required amount to assist pigment wetting, dispersion and paint stability is used – excess surfactant can have adverse effects on general paint properties such as paint stability, water sensitivity, wet adhesion, foaming tendency and the creation of other surface defects. Since choice can be very specific, reference should be made to the suppliers' literature to select the optimum for a particular application. However, in the case of decorative paints, reference has already been made to the common use of soya lecithin for use in alkyd-based products. It is also recognised that use of a lower viscosity polar medium for the dispersion stage may make the need for an additive less essential. In the case of latex paints a well proven basic dispersing system is a combination of a polyphosphate sequestering agent such as Calgon® or Tetron® with an organic dispersant such as Tamol 731® or equivalent from the Dispex® range. Typical levels would be 0.1% Tetron and 1.0% Tamol 731 on total pigment.

3.6.3 *Driers*

These materials are used to speed up the oxidation and polymerisation reactions in air-drying, oil-containing products. Normally soaps of heavy metals, they are usually employed as a mixture to control both surface and through drying. While various claims are made for the relative efficiency of the organic acids used in their preparation, it is the metal cation which acts as the catalyst in promoting faster dry.

Cobalt is the most powerful drier, but its effect is primarily at the surface. Excess of cobalt drier leads to premature surface skinning and subsequent wrinkling of the surface. A typical level of cobalt drier is 0.05% of cobalt metal, calculated on total resin solids.

Lead was the most commonly used drier to assist in 'through' drying, at a level of 0.5% on resin solids, but even this low level has come under criticism as presenting a possible toxic hazard, and now tends to be replaced by alternative through driers – though none has proven quite as effective as lead.

Manganese is also a powerful drier, contributing both to surface and through dry, but its colour tends to limit its use to primers and undercoats. The typical level of manganese usage is around 0.3%.

Calcium is finding increasing use as an auxiliary drier, at around 0.2% on resin solids in systems which have eliminated lead.

Zirconium is now regarded as the prime replacement for lead at typically 0.2% on resin solids, but its drying mechanism is rather different and it cannot be regarded as a direct substitution.

Aluminium is sometimes recommended as a potential part replacement for lead, but its use appears to be most effective when some modification of the resin is made to complement its effect. Reference to suppliers' recommendations is advised.

3.6.4 *Preservatives/biocides*

Aqueous-thinned products, particularly when containing a natural or synthetic colloid or cellulosic thickening agent, are liable to bio-degradation unless adequately protected with an effective preservative. Such protection should be afforded to any liquid raw material supplied or stored in bulk, as well as to the paint during manufacture and in the can.

Formaldehyde is commonly used to protect the latex in manufacture, delivery and storage, and can be a useful means of controlling, under carefully supervised conditions, the sterility of manufacturing equipment, pipelines, etc.

Being volatile, it is less effective as a long term 'in-can' preservative. A whole range of proprietary preservatives is available and reference to suppliers' literature will indicate optimum levels of use. It should be recognised, however, that this optimum level will be influenced by both the nature of the ingredients and the PVC of the formulation – higher PVC products tend to require higher levels of preservative for adequate protection.

It should also be stressed that use of 'in-can' preservatives is no substitute for good plant housekeeping – unless plant hygiene is maintained, problems of putrefaction or enzyme attack will be experienced, causing serious loss of viscosity. Even more important is the potential loss of production and product and the time taken to re-establish a safe manufacturing environment. The potential for such problems is far greater in situations where mercury-based preservatives cannot be used for safety reasons. No other type of preservative has been identified which has the same broad spectrum of activity as mercury-based products. Typical levels of use of in-can preservatives are 0.05–0.2% on total paint.

In formulating exterior coatings, it is necessary to consider the addition of film fungicides and algicides to prevent the growth of mould spores on the film surface. Again, a range of materials is available and care must be taken in the selection of grades appropriate to the type of product being modified. Higher levels of use are recommended than for 'in-can' preservatives, normally 0.5–1.0% on total paint. Often it is recommended that some zinc oxide is incorporated into the coating to give a boost to the level of protection afforded by the organic preservative.

3.6.5 *Insecticides*

Insecticides are not a common additive for paints, but from time to time interior paints are offered with insecticidal claims, particularly to help control flies. Other specialist uses are in marine applications to help control cockroaches. To be effective, fairly high levels of insecticide are required and the composition must be such that this can be 'fed' to the surface. Materials used have included Dieldrin®, Pyrethrum® and other chlorinated aromatic compounds which are claimed to have insecticidal properties.

3.6.6 *Can-corrosion inhibitors*

While fully lacquered cans are used for packaging aqueous paints, a corrosion inhibitor is highly desirable to prevent rusting from minor unprotected areas of tinplate. Sodium benzoate, at 1% on total water content, is the normal recommendation, although sodium nitrite is sometimes used as an alternative or in conjunction with the sodium benzoate.

3.6.7 *Flash rust inhibitors*

When aqueous paints are used directly on bare ferrous metal, it is quite common for discrete rust spots to spread into the film during the drying process, causing unsightly brown spots over the surface. Sodium benzoate or sodium nitrite at 1.5–2.0% on water content is often effective, but one or two proprietary products are available.

3.6.8 *UV absorbers*

These products are mainly of importance in 'clear-over base coat' automotive finishes; the use of approximately 1% of UV absorber improves the performance of the clear top coat. Used in conjunction with a 'light stabiliser', they reduce film degradation and overcome problems of under film chalking, delamination and cracking of the clear top coat. Proprietary products are available under the trade names of Tinuvin® and Sanduvor®.

3.6.9 *Reodorants*

A number of aromatic 'essential oils' have been suggested for masking the odour of paints during application and the immediate post-drying stage. In practice it is usually more acceptable to reduce as far as possible the cause of the odour by the elimination of trace monomers or by the use of low odour solvents. Possibly the most widely used additive in alkyd-based finishes is a low percentage of pine oil.

3.6.10 *Anti-skin*

One to two percent of pine oil or dipentine is often sufficient to minimise skinning of air-drying, solvent-thinned finishes in the can, but to give greater protection, particularly to avoid the formation of a very weak skin which cannot be removed and which would cause excessive bittiness in the finish, 0.2% of an oxime such as butyraldoxime or methyl ethyl ketoxime is commonly added.

3.6.11 *Anti-flood/float*

Certain combinations of pigments are much more prone to give lack of colour uniformity, due to selective flooding or floating, than others. Cogrinding and reduction in flocculation by ensuring good wetting and dispersion can reduce the problem but, when all else fails, the addition of an anti-float agent should be considered. Various proprietary products are available, but results can be very specific and should be checked carefully before arbitrary addition to a batch.

3.6.12 *Anti-foam*

The presence of surfactants in aqueous products is often responsible for the development of a stable foam which can persist through to the dry film. Use of an effective anti-foaming agent during the dispersion stage and as a post addition can reduce or eliminate this foam. Once again, a host of materials is available of varying types, ranging from mineral oils to alcohols, soaps or silicone-based materials or mixtures of these. The correct choice can be very specific and often cannot be confirmed in other than the practical conditions which cause the foam. It is a good basic rule to reduce as far as possible the development of foam, rather than having to correct it, by selection of low foaming surfactants and by attention to detail in the manufacturing operation.

3.6.13 *Anti-gassing*

Gassing in the can is usually caused by the presence of moisture in such products as moisture-cured polyurethane paints or products containing aluminium flake or powder or zinc dust. Good plant housekeeping should help to minimise sources of water contamination – often in the solvent – but the addition of a water-scavenging agent such as very fine particle silicas or additives based on monomeric isocyanate, normally at the pigment dispersion stage, is advisable to mop up any chance contamination.

3.6.14 *Anti-static*

Anti-static additives can be used at 0.3% level of addition to hydrocarbon solvents to reduce the possibility of electrostatic build up and hence potential risk of sparking and fire during storage and handling. A typical proprietary material is the anti-static additive ASA3 supplied by Shell Chemicals (UK) Ltd.

3.6.15 *Anti-freeze*

Aqueous products can be destabilised by the formation of ice crystals, which force the protective colloid or surfactants out of solution, leading to the resin partially coalescing in the can during the thawing process. Freezing point can be depressed by the addition of 1–2.0% ethylene glycol to the total formulation, but it is still advisable to recommend protecting such products from frost during storage.

Bibliography

M. Ash and I. Ash *Formulary of Paints and other Coatings Vol I* (1978), Vol II (1982), George Godwin Ltd., London.

E.W. Flick *Water-based Trade Paint Formulations* (1988); *Industrial Water-based Paint Formulations* (1988); *Contemporary Industrial Coatings. Environmentally Safe Formulations* (1985); *Interior Water-based Trade Paint Formulations* (1980); *Solvent-based Paint Formulations* (1977), Noyes Publications, Park Ridge, New Jersey, U.S.A.

D.A. Karsa (ed) *Additives for Water Based Coatings* Royal Society of Chemistry, London (1990).

R. Lambourne (ed) *Paint and Surface Coatings: Theory and Practice* Ellis Horwood Ltd., Chichester (1987).

R.R. Myers and J.S. Long (ed) *Treatise on Coatings, Vol 4 – Formulations*, Marcel Dekker, New York, 1975.

S. Paul *Surface Coatings Science and Technology* John Wiley & Sons, Chichester (1985).

H.F. Payne *Organic Coatings Technology, Vol I – Oils, resins, varnishes and polymers*, John Wiley, New York (1954)

O. Volk and M. Abriss *Federation Series on Coatings Technology—Unit 23 – Interior Finishes*, Federation of Societies for Coatings Technology, Philadelphia, 1976.

H. Warson *Gloss Emulsion Paints, 2nd edition* Solihull Chemical Services, Solihull, West Midlands (1983)

4 Volume relationships

FRED STIEG

4.1 Introduction

A considerable body of literature exists in technical journals describing methods for the determination of the 'critical' volume relationship between pigment and binder that is described as the critical pigment volume concentration or CPVC. Far less has been written about how this relationship, once identified, can be of assistance in the formulation of practical coatings.

Wolff [1–19], in Germany, first reported upon a 'critical oil content' in exterior oil paints that he described as a common transition point for the properties of blistering, rusting, and resistance to heat. His work was reviewed, and confirmed, by Elm in the USA [20], who shifted the emphasis from Wolff's 'critical oil content', by weight, to his own 'critical point in percent pigment/volume', although at that time the term PVC was already in fairly common use.

Some 13 years later, Thynne [21], in England, published a review of the pigment–binder relationship in which he described some original work involving the exposure of a series of metal primers in a salt-spray apparatus, reporting that '... degrees of rusting and blistering change inversely with percentage of pigment in the paint, and both curves show an abrupt change between 35 and 45 percent pigment'. He also referred to a discovery by Sherwin–Williams research that water vapor permeability (Payne cup method) exhibited a sharp increase at roughly the same percentage pigment by volume.

Of his own work Thynne commented, 'In industry today time does not permit the preparation of such graphs for each set of materials used, and a much quicker yet reliable method must be used to find this optimum combination.' Since this 'optimum', for his particular series, was in the semi-gloss range, he suggested that gloss might serve that purpose. Although the suggestion was naive, it would appear to be the first time that anyone had considered the problems that the working paint formulator might encounter in trying to apply the concept.

Any procedure that requires the laboratory preparation of a test series of finished paints, and identification of the CPVC by changes in a number of different physical properties as the PVC of the series is varied, still finds little application in industry today.

4.2 The CPVC concept

The existence of a critical pigment volume concentration is generally considered to have been discovered by Asbeck and Van Loo [22], of the Sherwin–Williams Company, despite its prior history briefly referred to above. They did much, however, to clarify the relationship by defining it in terms familiar to present-day paint formulators, and by offering a means for its determination that could be performed by the average paint laboratory in a relatively short period of time, since only one paint need be prepared.

The changes in dry-film properties described by Thynne [21] – rusting, blistering, vapor permeability, and gloss – were illustrated by a composite graph, from which Asbeck and Van Loo concluded that the CPVC was 'a transition point above or below which substantial differences in the appearance and behavior of a paint film will be encountered'.

They then defined the CPVC as 'that point in a pigment-binder system at which just sufficient binder is present to fill completely the voids left between the pigment particles after the volatilization of thinner'. In their discussion of this concept, however, Asbeck and Van Loo recognized that a portion of the binder was adsorbed on the pigment surface and was, therefore, distinct from that which merely filled the voids between particles.

They also theorized that the CPVC represented 'the densest packing of the pigment particles commensurate with the degree of dispersion of the system'. This being the case, they reasoned that it should be possible to determine the CPVC of a paint experimentally by filtering off any excess of vehicle, and measuring the apparent volume of the residual filter cake. For this purpose they used a modified Gooch 30 ml fritted-glass filter, calling it the 'Asbeck–SW CPVC cell'.

4.3 Establishing the CPVC

4.3.1 *The CPVC cell method*

The lid for the patented CPVC cell had a small filling hole in its center, to which a short piece of glass tubing was fused. The volume of the cell was first determined by smearing the fritted bottom of the cell with bodied linseed oil to prevent penetration of water, wiping off the excess, and filling the cell with water from a burette. The fritted bottom was then cleaned by drawing solvent through it, after connecting the cell to a suction flask.

Because most practical paint vehicles tend to filter very slowly, it was recommended that the paint to be tested was first diluted with an equal volume of suitable solvent; 10 ml of the diluted paint was then drawn into a syringe, and expressed into the cell. After drawing off the liquid portion of the sample by suction, the lid of the cell was replaced, and the cell refilled with water from a burette. The difference in volume between the two additions of water was the volume of the filter cake:

$$\text{CPVC} = \frac{\text{volume of pigment in sample}}{\text{volume of filter cake}}$$

In some instances, when a highly polymerized vehicle is relatively filter-proof, the developers of the method recommended that as much as possible of the original vehicle be first removed by repeated thinning with solvent, and centrifuging to separate the pigment.

Primarily because of results provided by the CPVC cell, Asbeck and Van Loo considered the CPVC to be influenced, among other factors, by the type of binder employed, and the fineness of grind of the system.

In a related paper, by Asbeck *et al.* [23], consideration was given to the relationship between pigmentation oil absorption and the CPVC, but the authors concluded that oil-absorption results were unreliable due to their lack of agreement with the results provided by the CPVC cell, and their relatively poor reproducibility. A later study has shown, however, that the reproducibility of the oil absorption test method can be greatly increased [24].

In a third paper, Asbeck, Scherer, and Van Loo described an 'ultimate pigment volume concentration' [25] that was obtained by tamping down the CPVC cell filter cake to obtain a smaller volume, assumed to represent more closely a condition of maximum packing. However, as noted by the authors, the CPVCs determined by this technique varied much less from those calculated from oil absorption than was the case in their earlier comparisons. This would seem to suggested that flocculation, possibly induced by their 'vehicle replacement' with thinner, which is required to permit the removal of excess binder by filtration, is responsible for the lack of agreement between the CPVC cell and oil absorption results, rather than variations in the latter.

Some confusion has also existed because the CPVC cell results were usually described as the 'CPVC of a paint' (or 'of a paint system'), as have the results of many other test procedures that have involved the evaluation of a series of paints of ascending PVC, while oil absorption results have been described as the 'pigmentation CPVC'. There can be no real difference, however, as the following argument will attest.

A 'paint' is a specific entity – a fixed combination of pigments, extenders, binders, solvents, additives, etc. – and, in industry, it is commonly the result of a prolonged trial-and-error process called, for lack of a better term, paint formulation. The use of a device such as the CPVC cell eliminates any need for a test series, but if all of the paints in such a series were pigmented with the same combination of pigment and extender (as they should be, since changes in the ratio of pigment to extender would alter their volume at maximum packing), and all were evaluated in the CPVC cell, all calculations would theoretically provide the same value for the CPVC.

While it would, of course, be possible to take such information, and formulate yet another paint at the indicated PVC, using the same ratio of pigment to extender, it would be sheer coincidence if the new product retained any of the characteristics of the original paint. As intended by Asbeck and Van Loo, the CPVC cell (under ideal conditions) establishes the lowest PVC at which the pigmentation of any tested oil paint will assume a condition of close packing, but only provides information as to whether the PVC of that paint is above or below the CPVC of its pigmentation. In other words, the CPVC is a *pigment* characteristic, rather than a paint characteristic.

There is of course a counter-argument. Several investigators [26, 27] have reported that CPVC, as determined by their methods, has varied for the same pigments dispersed in different vehicles, and for different degrees of dispersion in the same vehicle – presumably the effect of flocculation and/or agglomeration on pigment packing. Such variations have been cited as an indication that oil absorption was less reliable for establishing the CPVC, but are of less practical significance than suggested.

Some of these vehicle-to-vehicle variations of the CPVC have been induced by the use of such impractical test 'vehicles' as raw linseed oil, and a comparison with the results obtained using bodied or acid-refined oil [23]. One variation employed mineral oil to demonstrate, unrealistically, that flocculation affected results when using different vehicles. However, in his 1956 Mattiello Lecture, Van Loo [28] emphasized that the CPVC of any given pigmentation, as determined in the Asbeck–SW CPVC cell, did not vary for most practical paint vehicles.

It might seem more difficult to quarrel with Cole's [29] demonstration of the effect of dispersion on the CPVC (film density method), but the degree of dispersion in most practical paint products is today largely determined by the selection of pigments and extenders, and any unexpected variations from the standard set for fineness-of-grind are subject to correction in the mill room. Dispersion has not been found to be the source of any significant formulation problems, since most paint formulation actually represents the reformulation of an existing product possessing properties as close to those now desired as possible. The raw materials selected, and the production equipment employed, are those for which its manufacturer has already expressed a preference; their properties are already familiar and acceptable to the market served and fineness-of-grind is apt to be duplicated as well.

Interest in the CPVC cell method for determining CPVC has dwindled considerably in recent years, due to the increased production of latex paints. The particulate nature of a latex binder does not permit the excess to be removed by filtration.

4.3.2 *The oil absorption method*

The determination of pigmentation CPVC from the volume relationship between pigment and oil at the endpoint of the spatula rub-out oil absorption test [30, 31] had been in everyday use for controlling color and sheen uniformity in deep-tone alkyd flat wall paints for some time prior to the publication of an extensive series of two-component (TiO_2 and extender) CPVC curves [32] demonstrating the effects of pigment packing, and proving that oil absorption values were far from additive. This had been noted much earlier by Leisegang [33], and Klummp and Meier [34, 35].

Later work showed, however, that the CPVC for a three-component mixture was very close to the weighted average for two two-component mixes of the same titanium dioxide with each of the two extenders.

Table 4.1 Flat latex paint standard

	Original flat latex paint	
	kg	litre
TiO_2 83%	210.08	55.2
5.5 μm $CaCO_3$	288.12	106.2
Diatomaceous SiO_2	42.02	18.2
Latex 48%	366.15	345.8
Water	311.09	311.1
Additives	161.22	163.5
Totals	1378.68	1000.0

% PVC = 53.6	% Solids/vol = 33.5
% LP = 20.0	PVC/CPVC = 1.12
Cost = 1036 CU	HP/litre = 6.12 sq m

The possible variation in oil absorption results between operators was shown to be greatly reduced by the use of a 'driest possible' end-point, and still further by using a constant volume of 1.25 ml pigment (that of approximately 5 g of TiO_2 – a standard sample used in the titanium pigment industry control laboratories for many years) [36]. Using metric units in the calculation, the weight of each of the components in a blended pigmentation is multiplied by 1.25, and divided by the total pigmentation volume, to obtain the weight in grams of that component to use in a dry-blended sample for oil absorption (OA) test. A test sample for the pigmentation of Table 4.1 might be calculated as follows:

$$TiO_2\ 83\% \qquad \frac{210.08 \times 1.25}{179.6} = 1.4621\,\text{g}$$

$$5.5\ \mu\text{m}\ CaCO_3 \qquad \frac{288.12 \times 1.25}{179.6} = 2.0053\,\text{g}$$

$$\text{Diatomaceous}\ SiO_2 \qquad \frac{42.02 \times 1.25}{179.6} = 0.2924\,\text{g}$$

Due to the choice of sample size, the volumetric oil absorption of the blended pigmentation must be obtained by multiplying the millilitres of oil required to reach the described endpoint by 0.8. This is a simple procedure, and is a small price to pay to establish a practical constant volume for the test procedure.

The relationship of volumetric oil absorption to pigmentation CPVC is also simple [37, 38]:

$$\text{CPVC} = \frac{1}{(1 + \text{OA[v]})}$$

where OA[v] = volumetric oil absorption

Unlike the CPVC cell, volumetric oil absorption has proved useful for determining pigmentation CPVC in latex vehicles.

4.3.3 *Effect of latex vehicles on CPVC*

Because of its particulate nature, a latex binder cannot coat individual pigment surfaces with a thin film of adsorbed polymer, and the percentage of its particles capable of fitting into pigmentation voids will be a function of relative size. The overall effect is to create a discontinuity in the pigment dispersion, forming pockets of pigmentation [39], surrounded by a layer of unpigmented polymer, but has been attributed commonly to the reduced binding power of the latex.

Becker and Howell [40], in the USA, found that tensile strength, stain removal, scrub resistance, enamel hold-out, and weather–ometer performance all underwent an identifiable transition at a common PVC point that varied with latex particle size, plasticization, and glass transition temperature for any given pigmentation. The smaller the latex particles, and the greater their ability to deform at ambient temperatures, the higher the observed CPVC.

Berardi [41] suggested that the ratio of the volume of linseed oil required to wet and fill the voids of a given amount of pigment, to the volume of latex solids required to bind the same amount of pigment, should be termed the binding power index (BPI) of the latex vehicle.

Schaller [42] made observations for acrylic latex systems similar to those made by Becker and Howell for polyvinyl acetate, but also added the inception of high dry-hiding as an indication of CPVC. Adams, in England, had earlier related this hiding power transition point to the CPVC in oil paint systems [43].

Schaller's data, and his technique, were later used to develop a method for calculating numerical values for Berardi's BPI, making it possible to calculate the pigmentation CPVC, in a latex vehicle for which BPI was known, from oil absorption [44]:

$$\text{BPI} = \frac{(1 - \text{CPVC in oil})}{(1 - \text{CPVC in latex})}$$

$$\text{CPVC in latex} = \frac{1}{(1 + \text{OA[v]}/\text{BPI})}$$

4.3.4 *Other means for CPVC determination*

Because of the obvious unsuitability of the CPVC cell, and belief that oil absorption determinations could have no significance in systems that contained no oil, the producers of early latex polymer emulsions were responsible for adding a great many paint properties to the list of those identified by the original developers of the CPVC concept as undergoing a transition at the CPVC, quite a few of which have been mentioned already. In some instances, arbitrary levels of performance were selected as end-points, based upon transitions observed in other simultaneously tested properties.

With one exception, all of the test procedures had one feature in common that made them more suitable for research work than for practical paint formulation. They involved the preparation of a series of test paints covering a more- or less-extensive PVC range, and required numerous evaluations to determine a common transition point for all the properties selected. It was soon found that such properties did not all possess the 'common transition point' described in accepted definitions of the CPVC, some consistently suggesting higher CPVC values than others, and some exhibiting no transition at all [45].

In fact, the original common transition points could usually be so described only because properties such as the rusting of primed metal substrates, and the blistering of primers, are so obviously related to inverse aspects of water permeability. Even so, if the inception of blistering is taken as indicative of CPVC in one instance, and the inception of rusting in another, the two results will differ by quite a few PVC for the same pigmentation. The transition point is somewhere in the middle of the range, where both blistering and rusting are occurring simultaneously.

Such variations must be expected, because not all paint properties respond quickly to the increase in air-filled voids that must occur as the PVC of a test series is increased. Some, such as hiding power, respond to the presence of tiny discrete bubbles of air, while others, such as stain resistance and enamel hold-out, do not respond before enough air is present to form continuous pores. Still others, such as 'scrub' (or wet-abrasion resistance), do not appear to be related to the CPVC at all [46], the test having been devised for the comparative evaluation of latex vehicles for relative wet adhesion.

While it is thus impractical to expect optimum performance for all paint properties at any one PVC, this does not diminish the value of the CPVC, however determined, as a guide for paint formulation. Very few paints are actually formulated at their pigmentation CPVCs, since the CPVC seldom represents a point of optimum performance. The list of properties that will improve as the CPVC is exceeded (high dry hiding and blister-resistance) is far smaller than that of those that deteriorate (rusting, washability, gloss, levelling, chalk resistance, enamel hold-out, flexibility, resistance to efflorescence, and tensile strength), making less control of the relationship needed when formulating below the CPVC, but it can still be a useful benchmark.

Liberti and Pierrehumbert [47] produced the latex supplier's test procedure described as an exception at the beginning of this section because it did not require the production and evaluation of any series of test paints. Having arbitrarily decided that oil absorption values could not be applied to the calculation of CPVC for latex paints, because 'their dry films contain no oil' (and overlooking the fact that water also is not present), they attempted to replace oil absorption with 'water sorption'.

A 100 g sample of pigment was titrated with a 1.33% solution of equal parts by weight of Tamol® 731 and tripotassium polyphosphate, to an end-point at which 'the pigment slurry offered little or no resistance to stirring and flowed readily'. The volume of dispersant used (in ml) was then increased by 10% to obtain the pigment's water sorption. It was assumed that the water sorptions of pigments and extenders in a mixed pigmentation were additive, although oil absorptions are not, nor are packing densities.

This assumption made it possible to further simplify the method by publishing an extensive list of water-sorption values for commonly used pigments and extenders that could be employed in calculating the CPVC for practical latex paint formulations. For added convenience it was assumed that the CPVC would not be affected by the latex used, although their own results showed a variation of determined CPVC with latex particle size [48].

The procedure recommended was to divide the weight of each component of a mixed pigmentation (in pounds) by 100 and multiply by its water sorption. The gallons of vehicle solids required at the CPVC were then calculated by dividing the total of the water sorptions, obtained in the first step, by the density of the resin solids in pounds per gallon:

$$\mathrm{CPVC} = \frac{\text{volume of pigment}}{(\text{volume of pigment} + \text{vehicle solids})}$$

Despite the looseness of the end-point, the questionable assumption of additive water-sorption values, and calculation errors (the volume of resin solids would be the total of the water sorptions, divided by 8.33 – the density of water – not by 9.8), values of the CPVC obtained by this method were employed for years in the PVC/CPVC ratios used for the formulation of flat exterior latex house paints [49], which is a good illustration that the CPVC is not too critical a factor when formulating below it. At PVC/CPVC ratios greater than 1.00, however, it was found that high dry-hiding effects varied significantly for different pigmentations at the same PVC/CPVC ratio.

4.4 The relationship of PVC to CPVC

Among the first to utilize a relationship between the volume of binder required by a given pigmentation, and the volume supplied by the formula of which it was a part, to control paint properties, were Armstrong and Madson [50]. Their 'saturation value' for oil-based semi-gloss and flat wall paints was shown to represent a ratio affecting the properties of enamel hold-out, levelling, wet-abrasion, and vehicle penetration into a porous substrate:

$$\frac{\text{volume of (pigment+ binder)}}{\text{volume of (pigment+ binder) at saturation}}$$

This relationship can be expressed in terms of formula PVC and pigmentation CPVC (by the oil-absorption method). The conversion requires only the recognition that the volume of pigment+binder at the PVC is equal to pigment volume/PVC, and the volume of pigment+binder at saturation is equal to the same pigment volume/CPVC:

$$\text{saturation value} = \frac{\text{CPVC}}{\text{PVC}}$$

In attempting to interpret the meaning of this ratio, it is first necessary to recognize that the volume occupied by a paint film at its CPVC cannot shrink as less binder becomes available to fill the voids between pigment particles due to an increase in formula PVC. Air simply moves in to replace the binder, and the total volume remains the same as when saturated with binder.

Above the CPVC, therefore, the ratio of CPVC to PVC may be interpreted as:

$$\frac{\text{CPVC}}{\text{PVC}} = \frac{\text{volume of (pigment+ binder)}}{\text{volume of (pigment+ binder+ air)}}$$

Therefore

$$1 - \frac{\text{CPVC}}{\text{PVC}} = \frac{\text{volume of air}}{\text{volume of (pigment+ binder+ air)}}$$

$$= \text{decimal fraction in total dry paint film}$$

This will be recognized as identical to the common definition of porosity, as applied to a variety of materials, ranging from soils and rocks to cellulose derivatives. Consequently, if either of the ratios CPVC/PVC or PVC/CPVC is held constant, porosity will also remain constant.

Below the CPVC, the volume of binder at the formula PVC might be considered as the volume of binder required to saturate the pigment (at the CPVC) *plus* 'free binder', making the saturation value:

$$\frac{\text{volume of (pigment+ binder+ 'free binder')}}{\text{volume of (pigment+ binder)}}$$

Therefore

$$\frac{\text{CPVC}}{\text{PVC}} - 1 = \frac{\text{'free binder'}}{\text{volume of saturated pigment}}$$

and if either of the ratios CPVC/PVC or PVC/CPVC is held constant, the volume ratio of 'free binder' to the sum of pigment and fixed binder will also remain constant.

Considerable importance was attached by Vannoy [51, 52] to the concept of 'free binder' as a guide in formulating oil-base house paints of minimum oil content, although he used viscosity measurements for its identification, rather than the PVC/CPVC relationship. It has been shown, however, that interpretation of the Dintenfass [53] equation:

$$\log V[r] = \frac{F}{(100 - F)}$$

where

$$V[r] = \text{relative viscosity (dispersion/ vehicle)}$$
$$F = \text{\% sedimentation volume}$$

in terms of PVC and CPVC, will result in [54]:

$$\log V[r] = \text{\% solids} \times \frac{(\text{PVC/CPVC})}{(100 - \text{\% solids} \times \text{PVC/CPVC})}$$

Comparison with the original Dintenfass equation indicates that the PVC/CPVC ratio is equal to the percentage of sedimentation volume, divided by the percentage of solids/100, and since the percentage of solids by volume would be 100% for the oil vehicles used by Vannoy, and much earlier by Wolff [1–19], in exterior house paints seldom containing any volatile solvent, this may explain their finding a relationship between paint viscosity and durability.

This should not apply to latex paints, but some formulators [55] have used the ratio in producing good quality latex flats. It is noteworthy, however, that the paints formulated at the selected PVC/CPVC ratio (1.08) varied significantly in determined enamel hold-out and, even more significantly, in high dry-hiding.

4.5 The porosity concept

The importance of dry-film porosity can only be fully appreciated when it is recognized that, unlike the many paint characteristics that have been employed at one time or another for locating the CPVC, it is a *cause*, rather than a *symptom*, of over-all performance.

Porosity in a dry paint film is commonly recognized as the result of a deficiency of binder in the paint formula as compared to its pigmentation's binder demand. This definition necessarily deviates from the dictionary definition of porosity, as a general term, because the porosity of a paint would not be expected to be increased by the presence of *encapsulated* air, while the dictionary's definition makes no such distinction. Furthermore, a paint film would be considered completely porous in the absence of binder, which suggests that its relative porosity is better expressed by the relative volumes of air and binder in the voids of its closely packed pigmentation.

Since total void volume is equal to pigment volume multiplied by ((1− CPVC) / CPVC, and the volume of available binder to pigment volume multiplied by (1− PVC) / PVC, the decimal fraction of binder present will be (since the 'pigment volume' terms cancel out):

$$\frac{(1 - \mathrm{PVC})/\mathrm{PVC}}{(1 - \mathrm{CPVC})/\mathrm{CPVC}}$$

or

$$\frac{\mathrm{CPVC}(1 - \mathrm{PVC})}{\mathrm{PVC}(1 - \mathrm{CPVC})}$$

and since the fractions of binder and air add up to unity, the fraction of air in the *voids*, which is termed the porosity index (PI) [56], as distinct from the 'porosity' of Section 4.4, becomes:

$$\mathrm{PI} = \frac{1 - \mathrm{CPVC}(1 - \mathrm{PVC})}{\mathrm{PVC}(1 - \mathrm{CPVC})}$$

Since the volumetric oil absorption (OA[v]) is equal to 1/CPVC − 1 the equation may also be written as [57]:

$$\mathrm{PI} = \frac{1 - (1/\mathrm{PVC} - 1)}{\mathrm{OA[v]}}$$

$$1 - \mathrm{PI} = \frac{1/\mathrm{PVC} - 1}{\mathrm{OA[v]}}$$

It was shown experimentally that a logarithmic relationship existed between PI and vapor permeability (Payne cup), and a linear relationship between PI and dry hiding power, for a large number of tested solvent-type paints [58]. Paints formulated with similar PI values exhibit generally similar characteristics.

This was also found to be true for latex paints, but the point of zero porosity (which is the point at which the inception of high dry-hiding may be observed) occurred at a lower PVC in a latex vehicle than in an oil system, for the same pigmentation [40]. The relationship between this 'optical' CPVC for latex and the CPVC determined by oil absorption was employed to calculate Berardi's binding power index (BPI) as has previously been described [41]:

$$\mathrm{BPI} = \frac{\mathrm{OA[v]} \times \mathrm{CPVC}'}{1 - \mathrm{CPVC}'}$$

where

$$\mathrm{CPVC}' = \mathrm{CPVC}\ \text{in latex vehicle (optical)}$$

$$\mathrm{OA[v]} = \frac{1 - \mathrm{CPVC}}{\mathrm{CPVC}\ \text{in oil}}$$

This relationship also indicates that, for the same pigmentation, porosity in a latex system will vary with binding power index:

$$\frac{1 - \text{LP}}{\text{BPI}} = \frac{1/\text{PVC} - 1}{\text{OA}[v]} = 1 - \text{PI}$$

where

LP = latex porosity

The observed fact that a variety of paint properties other than optical porosity were affected by changes in BPI suggests that equal performance in these catagories might also be predicted from knowledge of either absolute or relative BPI values [59, 60]. The relative values for BPI remain constant, irrespective of the property employed to identify latex CPVC, which is to be expected because of their common relationship to dry-film porosity.

Casarini [61], in Italy, investigating this premise, has successfully demonstrated that a relative BPI value for two different latex systems (1.25 from her data), as determined from relative tensile strength evaluations, could be used to predict the PVCs at which equivalent scrub-resistance was obtained for a flat latex paint when switching from one vehicle to the other, or when changing pigmentations.

It has also been demonstrated that a plot of (1/PVC – 1) versus OA[v] for a given paint formulation may be employed to establish a porosity constant, identified in terms of (1 – LP) /BPI [36]. This is highly useful in rationalizing the ability to match latex paint porosities from oil absorption data, since the value of BPI merely alters the absolute value of porosity, but not the porosity match for any two paints made with the same latex vehicle. Any two paints falling upon a diagonal line extended from the origin of the plot will exhibit similar porosity (see Figure 4.1 for the example of Table 4.1).

4.5.1 *Effect of porosity on hiding power*

The hiding power developed by a given pigmentation is a function of the refractive index of the medium in which its particles are surrounded in a dry paint film. The greater the refractive index of the pigment, and the lower the refractive index of the medium, the greater the hiding power produced. This relationship was expressed by Fresnel in terms of relative 'reflectivity' [62]:

$$\text{Fresnel reflectivity} = \left(\frac{n[1] - n[2]}{n[1] + n[2]}\right)^2$$

where $n[1]$ = refractive index of pigment and
$n[2]$ = refractive index of medium.

In a porous paint film, that medium contains both air and binder, and its refractive index is a weighted average of the respective refractive indexes for

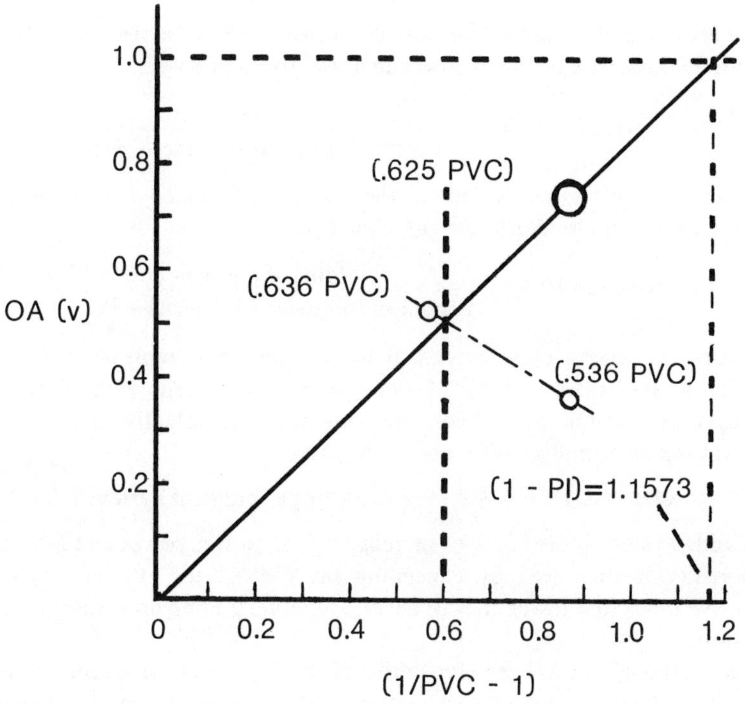

Figure 4.1 Locus of equal porosity.

air and binder (1.00 and 1.50), based upon the relative volumes of each present in the film.

It has already been noted in Section 4.5 that (1 – PI) and PI are equivalent to the fraction of binder and fraction of air, respectively, so the average $n[2]$ for the medium may be calculated as follows [63]:

$$\begin{aligned} n[2] &= \text{PI} \times 1.00 + (1 - \text{PI}) \times 1.50 \\ &= 1.50 - 0.5\,\text{PI} \end{aligned}$$

where $n[2]$ = average refractive index for medium

Obviously, as porosity increases, reflectivity will increase, but this only explains partly the well-known phenomenon of 'high dry-hiding' since porosity has yet another effect upon hiding power that is not quite so well known, or quite as obvious.

The hiding power produced by a given amount of titanium dioxide is a function of its volume concentration in the dry paint film in which it is dispersed [64, 65]. For a single-pigment, non-porous system, this concentration is equivalent to formula PVC, but as the PVC of the system is increased above the CPVC of its pigmentation, the total volume of the dry film is no longer predictable from its solids content [66]. The total volume is increased by the presence of air in the film, and the 'effective PVC' at which the titanium dioxide is dispersed becomes lower than the formula PVC:

$$\text{effective PVC} = \frac{\text{volume of pigment}}{\text{volume of (pigment + binder + air)}}$$

and since $(1 - \text{PI})$ is equivalent to the fraction of binder by volume in the mixture of air and binder in the pigment voids:

$$\text{effective PVC} = \frac{\text{volume of pigment}}{\text{volume of (pigment} + \text{binder}/(1 - \text{PI}))}$$

Finally, the above expression will be recognized as representative of the calculation associated with CPVC determination by means of the CPVC cell, and may be written as follows, showing that the effective PVC remains constant for all formula PVCs above the critical:

$$\text{effective PVC} = \text{CPVC (for a single-pigment system)}$$

This added dilution also acts to increase hiding power per pound of pigment as porosity is increased by exceeding the CPVC, the two effects (lower refractive index and lower effective PVC) together adding up to increased high dry-hiding.

Thus, although it has been shown that the hiding power produced per pound of titanium dioxide will decrease as formula PVC is increased, it can never be any lower than at its own CPVC. The concentration of titanium dioxide in any mixed pigmentation seldom, if ever, attains this level because of economic considerations, but the same statement may be made in respect to the CPVC of the combined pigmentation. The dilution of the titanium dioxide within the pigmentation at its CPVC, whatever it may be, is independent of binder content.

4.6 Use in formulation

Most practical formulation work in the paint laboratory involves *reformulation* of some product from the formula files, or preferably, one of the company's standard products, rather than the formulation of a completely new product from scratch. This is so for a number of very good reasons.

Most standard products represent the result of decisions made in the past as to raw materials, based upon performance, uniformity from one shipment to the next, cost, availability, and the service provided by their suppliers, not to mention personal preferences and prejudices involving sales personnel. Since some paint properties tend to be degraded by the same formula changes that improve some others, they also represent past compromises, based upon company and customer preferences as to the priorities accorded to such properties.

If the purpose of formulation work is the replacement of some such product, it will be obvious that a great amount of time may be saved by using the same raw materials and honoring the same preferences.

If the purpose of the work is the formulation of a new product for a new market, it is most probable that the prospective customer's desires (or the salesman's interpretation of his needs) will be couched in terms of similarities and differences in the properties and/or cost of some such familiar standard product. Most frequently, reduced cost or increased hiding power will be the primary objective.

If matching some competitive product is the goal, modification of the standard product that most closely approaches its properties remains the most promising approach.

4.6.1 *Spacing extenders*

The formulator's most useful tool in the process of product modification is a nodular extender of high-brightness, with a mean particle diameter of less than 1 μm (preferably 0.8 μm, or smaller). Such extenders have become commonly known as 'spacing extenders' or 'TiO_2 spacers', referring to their presumed ability to separate titanium dioxide particles in a dry paint film. Essentially, however, their function is that of replacing titanium dioxide volume without significantly affecting gloss or brightness, and of replacing binder volume without reducing hiding power.

The ability of any given extender to function in this manner has been quantified in terms of 'dilution efficiency' (E[d]), which was viewed as the ratio of its effective volume (as a TiO_2 diluent when calculating effective PVC) to its true volume, expressed as a percentage [67] (see Table 4.2 for typical values):

$$\text{effective PVC} = \frac{\text{volume of TiO}_2}{\text{volume of (TiO}_2 + \text{extender} \times \text{E[d]} + \text{binder)}}$$

It has also been speculated that E[d] values may represent a ratio between the volume percentage of a particle-size distribution that is fine enough to provide

Table 4.2 Particle size versus dilution efficiency

Extender type	E[d] value
12.5 μm $CaCO_3$	0
7.5 μm talc	0.07
5.5 μm $CaCO_3$	0.15
3.0 μm $CaCO_3$	0.30
2.0 μm SiO_2	0.40
1.8 μm Calcined clay	0.74
1.0 μm Calcined clay	0.85
0.8 μm $CaCO_3$	0.99

dilution, and that which can merely produce a 'window through the film' effect [68] because of its size in relation to normal dry-film thicknesses.

4.6.2 *Modifying a standard product*

When a standard product is to be replaced by a modified version to reduce its cost, or its content of some raw material that is in short supply, or excessively high-priced, the most important consideration is that of retaining as many of the product's original combination of properties as possible. Regardless of product type, one of the most important of these properties will be tinctorial strength, to avoid problems relating to any standard color system normally in use.

Maintaining the same hiding power in a gloss or semi-gloss formulation, while reducing titanium content, is most easily accomplished by increasing the percentage of total solids by volume (which will increase hiding power in direct proportion to the increase), and by then replacing titanium dioxide with a spacing extender, on an equal-volume basis, until the original hiding power is regained [69]. As a general rule of thumb, the percentage of titanium dioxide that may be replaced will be approximately *twice* the desired percentage decrease in hiding.

Table 4.3 Standard for hiding power

	Original latex semi-gloss	
	kg	litre
TiO_2 95%	288.12	70.4
2.0 μm SiO_2	30.01	11.3
Latex 48%	600.24	561.8
Water	112.69	112.7
Additives	246.34	243.8
Totals	1277.40	1000.0
% PVC = 24.7	% Solids/vol = 33.1	
Cost = 1433 CU	HP/litre = 7.32 m^2	

In the case of a semi-gloss formulation such as that shown in Table 4.3 [70], the flatting extender (silica) in the original formula would be retained as the same percentage of total solids volume, to avoid an undesirable increase in gloss. If it is assumed that a 10% increase in total solids is desired, all components, other than the additives and water, are increased by 10% [71], and the volume re-adjusted to 1000 litres by removing water (Table 4.4). Since this formulation would develop 10% more hiding power than required, the next step will be to calculate a second blending base by replacing 25% of the titanium pigment with an equal volume of spacing extender (following the rule-of-thumb). A matching blend (Table 4.5) will contain 13% less titanium pigment per gallon than the original, and cost slightly less.

Table 4.4 Increased total solids

	Modified latex semi-gloss	
	kg	litre
TiO_2 95%	316.93	77.4
2.0 μm SiO_2	27.50	12.5
Latex 48%	660.26	618.0
Water	48.34	48.3
Additives	246.34	243.8
Totals	1304.87	1000.0
% PVC = 24.7	% Solids/vol = 36.5	
Cost = 1562 CU	HP/litre = 8.06 m^2	

Table 4.5 Hiding power match at higher solids

	Modified latex semi-gloss	
	kg	litre
TiO_2 95%	249.75	61.0
0.7 μm $CaCO_3$	44.56	16.4
2.0 μm SiO_2	33.01	12.5
Latex 48%	660.26	618.0
Water	48.30	48.3
Additives	246.34	243.8
Totals	1282.22	1000.0
% PVC = 24.7	% Solids/vol = 36.5	
Cost = 1426 CU	HP/litre = 7.32 m^2	

Calculating a hiding power match at higher total solids by computer requires only a single operation, and involves incremental steps that are only practical for a computer [72]. In this particular case, the match contains about 20% less TiO_2 than the first blending base.

4.6.3 *Latex paints*

As long as the same latex vehicle is employed in any modification of a standard product, its binding power index (BPI) need not be known to obtain a satisfactory match in properties using the procedure of Section 4.6.2. However, if a vehicle change is involved in the reformulation of a porous flat wall paint, any change in BPI must be taken into account, because both hiding power and porosity should be matched simultaneously for a true product-match.

Absolute values for BPI are not required. It has been found that a relative BPI will serve the purpose [61], and that it may be determined most accurately from a comparison of the PVC levels at which high dry-hiding is first observed, using Schaller's method for a single extender [62]:

$$\mathrm{BPI}[a/b] = \frac{1/\mathrm{PVC}[a] - 1}{1/\mathrm{PVC}[b] - 1}$$

where

$$\begin{aligned} \text{BPI[a/b]} &= \text{relative BPI (latex A/latex B);} \\ \text{PVC[a]} &= \text{apparent CPVC}', \text{latex A;} \\ \text{PVC[b]} &= \text{apparent CPVC}', \text{latex B.} \end{aligned}$$

4.6.4 *Flat latex paints*

The flat latex paint of Table 4.1 will be used as an example of the calculations involved in matching both hiding power and porosity after a pigmentation has been modified by replacing a latex-grade titanium dioxide with an enamel grade, plus spacing extender. Using the same latex vehicle, the original 210.08 kg of 83% titanium pigment was replaced with a 95% grade as follows:

$$210.08 \times \frac{83}{95} = 183.54$$

TiO_2 83%	210.08 kg	55.2 litre
TiO_2 95%	183.54 kg	44.8 litre
	26.54 kg	10.4 litre

$$10.4 \times 2.7 = 28.08 \text{ kg} \qquad 0.7\ \mu\text{m } CaCO_3$$

Because the actual titanium dioxide content has not been changed, and its effective PVC remains the same, the hiding power of the modification shown in Table 4.6 would be the same as that of the original if the porosities of the two formulations were identical, but the replacement of the high-oil-absorption TiO_2 with a combination of low-oil-absorption TiO_2 and extender will obviously have decreased porosity, and high dry-hiding along with it. The PVC must be raised.

Table 4.6 Effect of titanium pigment replacement

	Modified flat latex paint	
	kg	litre
TiO_2 95%	183.54	44.8
0.7 μm $CaCO_3$	28.13	10.4
5.5 μm $CaCO_3$	288.12	106.2
Diatamaceous SiO_2	42.02	18.2
Latex 48%	366.15	345.8
Water	311.09	311.1
Additives	161.22	163.5
Totals	1380.27	1000.0
% PVC = 53.6	% Solids/vol = 33.5	
% LP = 0	PVC/CPVC = 0.82	
Cost = 983 CU	HP/litre = 4.99 m^2	

The first step is to determine the porosity of the original, to be used as a standard. This is done by calculating the dry blend of pigments to be used for the volumetric oil absorption determination using the procedure described in Section 4.3.2. For this combination, the test end-point required 0.935 ml of oil, corresponding to an OA[v] of 0.748. Therefore, as illustrated in Figure 4.1:

$$\frac{1-\text{LP}}{\text{BPI}}=\frac{1/0.536-1}{0.748}$$

$$=1.1573$$

For the re-pigmented formula at 0.536 PVC (Table 4.6), the required volume of oil amounted to only 0.459 ml, and OA[v] to 0.367.

Table 4.7 Effect of 10% PVC increase

	Modified flat latex paint	
	kg	litre
TiO_2 95%	183.54	44.9
0.7 μm $CaCO_3$	119.06	43.9
5.5 μm $CaCO_3$	288.12	106.2
Diatomaceous SiO_2	42.02	18.2
Latex 48%	287.37	271.4
Water	351.90	351.9
Additives	161.22	163.5
Totals	1433.23	1000.0
%PVC = 63.6	% Solids/vol = 33.5	
%LP = 26.0	PVC/CPVC = 1.13	
Cost = 925 CU	HP/litre = 6.46 m^2	

The next step is to calculate the pigmentation for an arbitrary 10 PVC increase, obtained by replacing vehicle solids with spacing extender (Table 4.7). Again basic hiding power will not have been changed. For this combination, OA[v] = 0.536, and the equation for a line connecting the two points is:

$$\frac{1/0.536-1/0.636}{0.536-0.536}=1.7358$$

Therefore, for the desired porosity match:

$$\frac{1/\text{PVC}-1/0.636}{0.536-\text{OA[v]}}=1.7358$$

But for the locus of equal porosity:

$$1/\text{PVC}-1=1.1573\times\text{OA[v]}$$

Solving the equations simultaneously for OA[v], PVC = 0.625, and at this PVC, hiding power and porosity will match (see Table 4.8). Since the same latex vehicle was used, BPI has no effect on the match.

The triangular graph of Figure 4.2 illustrates the relationships involved [75]. All possible formulations possessing a latex porosity of 20% must fall on the straight line connecting the *pigment* apex of the graph with a point 20% of the distance between the *binder* apex and the *air* apex. Similarly, the positions of all possible formulations pigmented at the PVCs of 53.6% and 62.5% may be located by straight lines drawn from the *air* apex to the appropriate points on the base of the graph. It will be noted that porosity represented by the ratio (1 – CPVC)/PVC (fraction of air in total dry film) does *not* remain constant for a constant value of LP. This provides an explanation for the previously mentioned dry-hiding power variations observed when attempts have been made to base reformulation upon a constant PVC/CPVC ratio.

Table 4.8 Hiding porosity match

	Modified flat latex paint	
	kg	litre
TiO_2 95%	183.54	44.8
0.7 μm $CaCO_3$	108.62	40.0
5.5 μm $CaCO_3$	288.12	106.2
Diatomaceous SiO_2	42.02	18.2
Latex 48%	296.25	279.8
Water	347.45	347.5
Additives	161.22	163.5
Totals	1427.22	1000.0
%PVC = 62.5	% Solids/vol = 33.5	
%LP = 20.0	PVC/CPVC = 1.09	
Cost = 935 CU	HP/litre = 6.12 m^2	

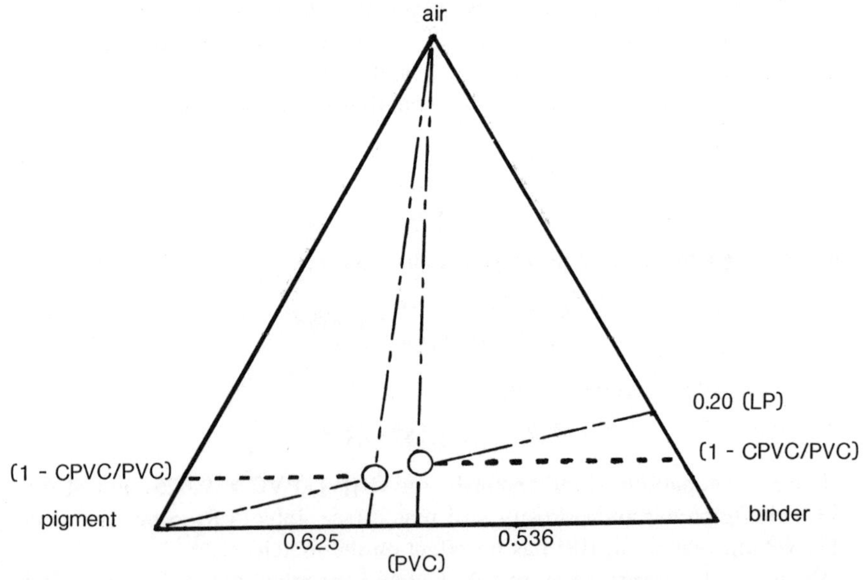

Figure 4.2 Relationship of LP and common porosity.

4.6.5 *Effect of relative BPI*

If it had been necessary to change the latex vehicle to one for which BPI[r] (relative to the original) was determined to be 0.8, then the value of (1 – LP) / BPI for the original formulation, using this new vehicle, would be identified as:

$$\frac{1.1573}{0.8} = 1.4466$$

making the equation for the original:

$$(1/\text{PVC} - 1) = 1.4466\,\text{OA[v]}$$

Now when the two equations are solved simultaneously for OA[v]

$$\%\ \text{PVC} = 59.4$$

for the porosity match (Table 4.9).

Table 4.9 Hiding porosity match with new latex

	Modified flat latex paint	
	kg	litre
TiO_2 95%	183.54	44.8
0.7 μm $CaCO_3$	80.31	29.6
5.5μm $CaCO_3$	288.12	106.2
Diatomaceous SiO_2	42.02	18.2
Latex 48%	320.83	303.0
Water	334.66	334.7
Additives	161.22	163.5
Totals	1410.70	1000.0
%PVC = 59.4	% Solids/vol = 33.5	
%LP = 20.0	PVC/CPVC = 1.10	
Cost = 951 CU	HP/ litre = 6.12 m^2	

It should be noted that if the original latex paint is made up with this new vehicle, and again matched, the PVC of matching porosity and hiding power is found to be 0.625, the same as before. However, the levels of both hiding power and porosity are higher.

4.6.6 *Matching a competitive sample*

A competitive sample of unknown composition may be matched for hiding power in this manner, but not for porosity. It is therefore necessary to select a product of known composition that exhibits adequately close dry-film porosity, as judged by stain-removal or enamel hold-out.

The hiding power of the unknown, relative to the selected formula, may then be determined by tinting equal-volume aliquots with exactly the same volume of dispersed colorant, and measuring dry-film reflectance for each [73]:

$$\text{relative HP}[u/s] = (1 - R[s]) / (1 - R[u])^2 \times R[u]/ R[s]$$
where $R[s]$ = reflectance of standard
$R[u]$ = reflectance of unknown

The hiding power of the standard may then be raised or lowered to provide a match by multiplying the volume of all solid constituents by the value obtained for HP[u/s]. This will not change dry-film porosity, and provides a new standard.

If the value of HP[u/s] is less than 1.00, a hiding power match containing less TiO_2, and costing somewhat less, may be obtained by replacing TiO_2 with spacing extender, rather than lowering the percentage of total solids, but as indicated in Section 4.6.2, this requires an incremental approach best handled by computer, and readjustment of the PVC for a porosity match.

If the value of HP[u/s] is greater than 1.00, and cost considerations tend to rule out an increase in the percentage of total solids, additional hiding power at equal porosity may be obtained by replacing some of the extender present with a finer particle size material. If this should increase total pigmentation oil absorption, as is likely, a reduction in PVC may be required for a porosity match, again increasing cost.

In this event, it will be necessary to consider the possibility that the product originally selected for modification was actually somewhat less porous that the competitive product, instead of a close match. In the case of the formula used as a standard for this exercise, the porosity was quite low (LP = 0.20), and an increase would be desirable to reduce wet-to-dry hiding power loss.

A comparison for relative wet-to-dry hiding power change with the competitive sample can be informative. This is best accomplished by the tinctorial strength method previously described, adding exactly the same volume of dispersed colorant to equal volumes of each.

All tinted latex paints will darken as they dry – due to the effect of water evaporation on the refractive index of the vehicle portion, and the accompanying film shrinkage that increases TiO_2 effective PVC – and reach a point of minimum reflectance when their pigmentations have reached a condition of maximum packing [74]. Generally speaking, the increase in reflectance that takes place *after this minimum has been reached* will be proportional to dry-film porosity. The total wet-to-dry change is not a reliable indication, because the percentage of total solids volume may differ for the two formulations.

If this comparison suggests greater porosity for the competitive product, porosity may be increased either by replacing some of the extender with a high-oil-absorption type, or by increasing formula PVC. If the latter, a fairly accurate match for the unknown's hiding power may be obtained by repeating the calculations of Section 4.6.4, assuming a reasonable value for BPI (such as 0.69) so that LP may be calculated. Then increase LP by HP[u/s] – 1:

$$LP = 1 - 1.1573 \times 0.69 = 0.2015$$

$$LP' = 0.2015 + HP[u/s] - 1$$

as before

$$\frac{1/PVC - 1/0.636}{0.536 - OA[v]} = 1.7358$$

but now

$$(1/PVC - 1) = (0.2015 + HP[u/s]) \times OA[v]$$

In this procedure, the adjustment was based upon the observation that the increase in high dry-hiding for a given increase in porosity is approximately a linear relationship [65]. Thus it is unnecessary to repeat any OA[v] determinations, and the match for porosity is not affected by the value assumed for BPI. It is convenient, therefore, if not too precise for large increases in high dry-hiding.

Some other alternatives are discussed in Chapter 5 and if increased porosity is undesirable, the use of opaque polymer is suggested.

4.7 Summary

Evidence has been overwhelming that no other factor exerts as great an influence upon the performance of a pigmented coating as the relationship existing between the volume of binder supplied by its formula and the volume of binder required to satisfy the binder-demand of its pigmentation. This latter factor is easily determined as the volumetric oil absorption of the dry-blended pigmentation, using the spatula rub-out method, and is just as meaningful in latex vehicles as in solvent-base systems, a point that has been emphasized in this chapter by the use of latex paint examples.

References

1. H. Wolff, *Farben-Ztg.* **35** 1031 (1930).
2. H. Wolff, *Farben-Ztg.* **35** 2175 (1930).
3. H. Wolff, *Farbe Chem.* **2** 297 (1931).
4. H. Wolff, *Oberflachtech.* **10** 156 (1932).
5. H. Wolff, *Farben-Ztg.* **37** 876 (1932).
6. H. Wolff, *Farben-Ztg.* **37** 1512 (1932).
7. H. Wolff, *Korrosion Metallschutz* **9** 57 (1933).
8. H. Wolff, *Farben-Ztg.* **38** 876 (1933).
9. H. Wolff, *Paint Varnish Production Manager* **9** 5 (1933).
10. H. Wolff and G. Zeidler, *Kolloid-Z.* **65** 228 (1933).
11. H. Wolff and G. Zeidler, *Farben-Ztg.* **38** 1495 (1933).
12. H. Wolff and G. Zeidler, *Paint Varnish Production Manager* **9** 18 (1933).
13. H. Wolff and G. Zeidler, *Korrosion Metallschutz* **10** 157 (1934).
14. H. Wolff and G. Zeidler, *Farben-Ztg.* **40** 375 (1935).
15. H. Wolff and G. Zeidler, *Paint Varnish Production Manager* **12** 14 (1935).
16. H. Wolff and G. Zeidler, *Kolloid-Z.* **74** 97 (1936).
17. H. Wolff and G. Zeidler, *Paint Varnish Production Manager* **17** 7 (1937).
18. H. Wolff and H. Hesse, *Farben-Ztg.* **42** 1062 (1937).
19. H. Wolff and G. Zeidler, *Farben-Ztg.* **42** 357 (1937).

20. A. C. Elm, *Ind. Eng. Chem.* **26** 1245 (1934).
21. A. W. F. Thynne, *Paint Technol.* **11** 423 (1946).
22. W. K. Asbeck and M. Van Loo, *Ind. Eng. Chem.* **41** 1470 (1949).
23. W. K. Asbeck, Laiderman and M. Van Loo, *Offic. Dig.* **24** 156 (1952).
24. F. B. Stieg, *American Paint J.* **43** (4) 106 (1958).
25. W. K. Asbeck *et al*, *Ind. Eng. Chem.* **47** 1472 (1955).
26. F. M. Smith and D. M. Stead, *J.O.C.C.A.* **37** 194 (1954).
27. G. G. Sward, *Paint Testing Manual* (13th Ed.) 239 (1972).
28. M. Van Loo, *Offic. Dig.* **28** 1126 (1956).
29. R. J. Cole, *J.O.C.C.A.* **45** 776 (1962).
30. F. B. Stieg and D. F. Burns, *Offic. Dig.* **26** 81 (1954).
31. F. B. Stieg, *Offic. Dig.* **28** 695 (1956).
32. *CPVC of Various Pigment-Extender Combinations*, NL Industries (1954).
33. R. E. Leisengang, *Farben-Ztg.* **33** 2424 (1928).
34. E. Klummp and E. Meier, *Farbe Lack* **34** 127 (1929).
35. E. Klummp and E. Meier, *Farbe Lack* **34** 599 (1929).
36. R. H. Rowland and F. B. Stieg, *J. Coatings Technol.* **54** 51 (1982).
37. F. B. Stieg, *J. Paint Technol.* **39** 701 (1967).
38. *CPVC of Various Pigment-Extender Combinations*, NL Industries (1959).
39. F. B. Stieg, *Offic. Dig.* **25** 226 (1953).
40. J. C. Becker and D. D. Howell, *Offic. Dig.* **28** 775 (1956).
41. P. Berardi, *Paint Technol.* **27** 24 (1963).
42. E. J. Schaller, *J. Paint Technol.* **40** 433 (1968).
43. P. I. Adams, *J.O.C.C.A.* 295 (1961).
44. F. B. Stieg, *J. Paint Technol.* **42** 329 (1970).
45. G. P. Bierwagen, *J. Paint Technol.* **44** 46 (1972).
46. C. C. Tatman, *J. Coatings Technol.* **50** 64 (1978).
47. F. P. Liberti and R. C. Pierrehumbert, *American Paint J.* **42** 22 (1958).
48. R. C. Pierrehumbert and C. Boyce, *Can. Paint Varnish Mag.* **39** 59 (1965).
49. R. C. Pierrehumbert, *American Paint J.* **68** 49 (1964).
50. W. G. Armstrong and W. H. Madson, *Offic. Dig.* **19** 321 (1947).
51. W. G. Vannoy, *Offic. Dig.* **21** 235 (1949).
52. W. G. Vannoy, *Offic. Dig.* **33** 1215 (1961).
53. L. Dintenfass, *J.O.C.C.A.* **41** 125 (1958).
54. F. B. Stieg, *Offic. Dig.* **31** 736 (1959).
55. A. Ramig Jr., *J. Paint Technol.* **47** 60 (1975).
56. F. B. Stieg, *Offic. Dig.* **34** 1065 (1962).
57. F. B. Stieg, *American Paint J.* **53** 46 (1978).
58. F. B. Stieg, *Offic. Dig.* **33** 792 (1961).
59. F. B. Stieg, *J. Paint Technol.* **42** 329 (1970).
60. F. B. Stieg, *ACS Organic Coatings and Plastics Chem.* **36** (2) 443 (1976).
61. A. Casarini, *Paint Mfr.* 10 (April 1973).
62. A. J. Fresnel, *Annal Chimie et Physic* **17** 312 (1821).
63. F. B. Stieg, *J.O.C.C.A.* **53** (6) 469 (1970).
64. F. B. Stieg, *La Chemie des Peintures* **20** (9) 353 (1957).
65. F. B. Stieg, *I&EC Product R&D.* **13** 41 (1974).
66. P. E. Pierce and R. M. Holsworth, *Off. Dig.* **37** 272 (1965).
67. F. B. Stieg, *J. Coatings Technol.* **53** 75 (1981).
68. F. B. Stieg, *J. Coatings Technol.* **61** 67 (1989).
69. F. B. Stieg, *American Paint and Coatings J.* **74** (23) 36 (1989).
70. Rohm and Haas Company *Formulation SG-41-1*, October 1978.
71. P. B. Mitton, *Offic. Dig.* **29** (1957).
72. PigmenTech Consulting, 903 Beachview Dr., Jekyll Is., GA 31520 USA.
73. E. Lund and F. Vial, *Farbe Lack* **65** 695 (1959).
74. F. B. Stieg, *Progress in Organ. Coatings* **1** (4) 351 (1973).
75. F. B. Stieg, *J. Coatings Technol.* **55** 111 (1983).

5 Dry hiding – an alternative to titanium dioxide

RICHARD WOODBRIDGE

5.1 Introduction

In the search for more cost effective decorative paints, especially since the early 1970s when most raw material and particularly TiO_2 prices escalated at a far faster rate than the price of finished products, attempts have been made to reduce the concentration of titanium dioxide without sacrifice of dry hiding. Reference has been made earlier to the titanium dioxide normally contributing most to the cost of the finished product and any savings in this area are well worthwhile.

Table 5.1 Indicative costs of raw materials on solids volume basis

	Indicative price (1000 kg)	Specific gravity (solids)	Cost/kg (solids)	Cost/l (solids)
Titanium dioxide	1500	4.0	1.5	6.0
Acrylic latex binder	1240	1.12	2.96	3.31
Vinyl latex binder	450	1.17	0.86	1.0
Extender	100	2.6	0.10	0.26

Table 5.1 shows the significance of calculating the cost of main paint ingredients on a volume basis, and how this highlights the impact of the much higher specific gravity of the titanium dioxide, compared with most other paint constituents, on the relative costs of key materials. Taking this a stage further, the typical volume relationships of key ingredients in a range of latex-based paints from 15–90% pigment volume concentration (PVC) can be studied as shown in Figure 5.1. Applying the cost data from Table 5.1, the relative cost of these key ingredients over the same PVC range may be achieved, as shown in Figure 5.2. The cost contribution of TiO_2 exceeds the cost of other ingredients once 30% PVC is exceeded, with a peak at around 35% PVC – the optimum level of pigmentation for semi-gloss latex paints which dominate the decorative latex paint market. It can be seen that there is considerable economic benefit if the required level of opacity and other desired properties are obtained by the use of minimum quantities of titanium dioxide.

In white paints, opacity is caused by the scattering of incident light and is a function of the difference in refractive index between pigment and binder. The proportion of incident monochromatic light reflected by a surface

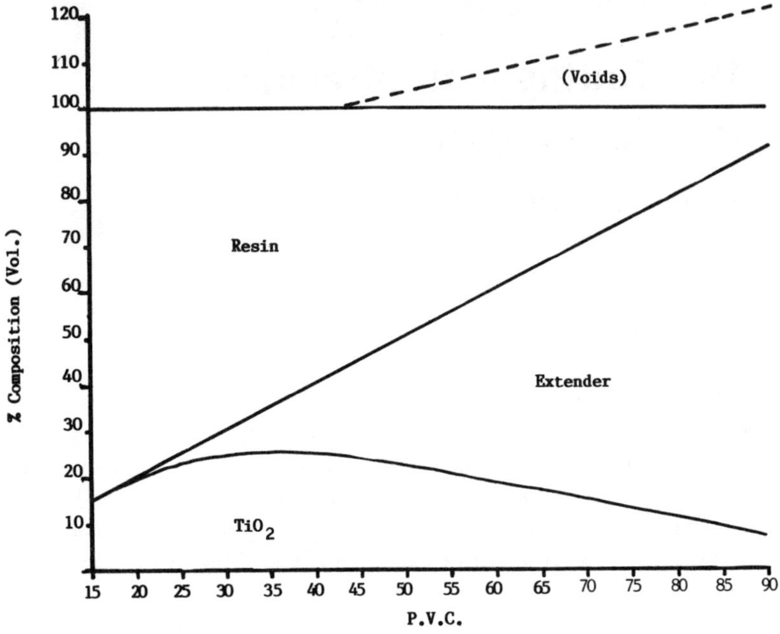

Figure 5.1 Volume relationships in latex paints.

perpendicular to the incident beam (I_0) is given by Fresnel's Law of Reflection:

$$I_R = \left(\frac{n_1 - n_2}{n_1 + n_2}\right)^2 I_0$$

where n_1 and n_2 are the refractive indices of the media involved. Rutile titanium dioxide is an effective white pigment because of the difference in its high refractive index (2.71) from that of the binder (usually about 1.5). Voids in a paint film contribute to opacity because the refractive index of air (1.0) is again different from that of the binder.

The degree of scattering by either a pigment particle or a microvoid is also influenced by the spacing of the scattering sites and their size. For maximum diffraction the scattering sites should be separated from each other by one site diameter, since closer packing will produce overlapping diffraction fields and wave interference. It has been calculated for pigments that the particle diameter should be one half of the wavelength of the light being scattered. Titanium dioxide should have an optimum particle diameter of 0.25–0.3 μm to scatter white light with an average wavelength of 5000 Å (0.5 μm).

The optical properties of microvoid systems, with and without other pigmentation, have been studied by a number of workers. The definitive work in this area is due to Kerker and Cook (1976) and Seiner (1977).

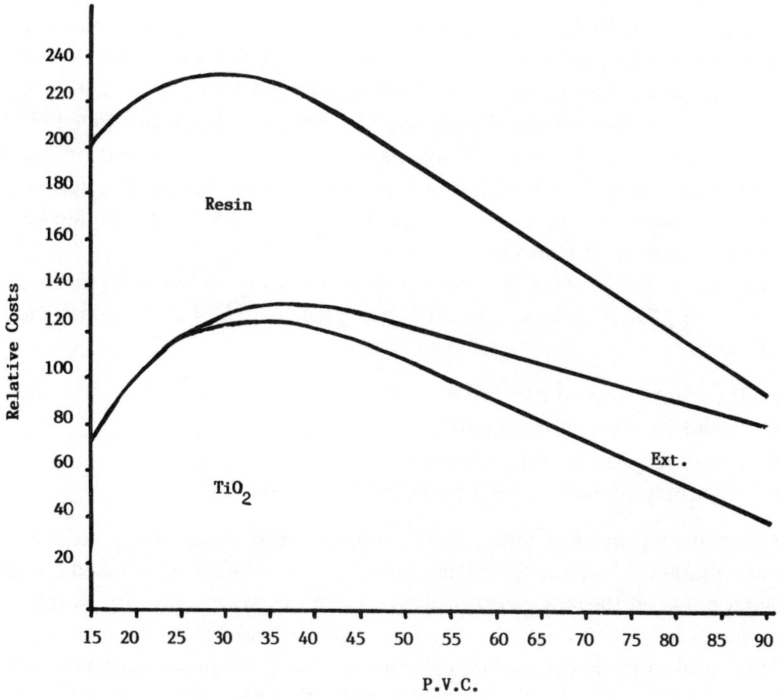

Figure 5.2 Relative costs of major raw materials in latex paints versus PVC.

It is concluded that for pigment-free microvoid systems of perfectly uniform geometry, optimised scattering is attained when microvoid size is about one third of the spacing between microvoids. For systems containing separate microvoids and titanium dioxide particles, and where at least partial flocculation or agglomeration exists, the results suggest the possibility of much improved hiding or much reduced prime pigment content.

The objective of this chapter is to discuss the various methods for the controlled incorporation of voids. Most of these methods are subject to patent claims and recognition of this is important if the technologist considers using any of the techniques discussed. While recognising that certain techniques may be applied to solvent-thinned products, the major use of this microvoid technology is in the field of latex paints, and it is this aspect which will be considered in detail.

5.2 Available technologies

The concept of achieving hiding by having an air/pigment interface, instead of having the pigment totally bound by the binder, is not new. So-called 'ceiling white' or 'dry distemper', based on whiting and glue solution, used this principle for many years. Such coatings had virtually no wash resistance and their use was limited to surfaces such as ceilings where minimum performance requirements existed. However, the concept of formulating latex paints above their critical pigment volume concentration (CPVC) to achieve a degree of dry hiding without undue loss of performance has been developed significantly over the last 30 years. Many papers have been published on the role and mechanisms of use of different types or sizes of extenders, often contradictory, but with one common aspect: all such paints, being formulated above the CPVC, give matt finishes.

Alternative routes must be found if it is necessary to build dry hiding into silk or semigloss finishes. Such technologies, which will be considered in detail, are:

- Pittment – closed cell microvoids.
- Spindrift – vesiculated particles.
- Plastic pigment – polymeric extender.
- Opaque polymer – single particle microvoids.

Before considering these individually, some general observations are relevant. For any market or market sector there normally will be an established optimum balance between the key paint properties such as sheen, opacity, scrub/wash resistance, durability, cost, etc. This may put a constraint on the use of certain technologies in some product types while offering real opportunities in others.

In examining an alternative technology the possibility of achieving a particular balance of properties not attainable with other techniques, which could lead to new market opportunities, should be recognised as should the fact that all techniques available for air entrainment as a means of achieving opacity, result in a reduction in wet hiding compared with more conventional formulations of equivalent dry hiding. This may present a problem in products designed for use by the professional decorator, since his attempt to maintain the normal level of 'wet' hide will result in a reduction in spreading rate. Clearly the improvement in hiding on drying is a benefit but education is needed. For the DIY market, this benefit of increased hiding on drying is certainly an advantage.

In considering the use of alternative technologies within world markets, considerable differences can exist in the potential cost benefits of one technology, relative to another, depending on local differences in costs of key components. This can particularly affect the relative costs of the various polymer-based technologies. Even within one country significant differences in the cost of one polymeric extender against another may be found, depending on whether it has to be bought in or is of 'own' manufacture.

A further factor influencing the potential for any of these technologies is the degree of formula optimisation already achieved. The greater the existing cost optimisation of any formulation, the more difficult it will be to achieve a saving using alternative technology, unless that technology is also fully optimised.

All of these factors highlight the need to assess carefully the relevance and potential benefits of each alternative technology against specific product requirements. It should also be clear from these comments that something which works well in one formulation may offer no benefits whatsoever in another, despite being superficially similar in general properties. Advantage of the options available can only be taken by learning *how* to optimise each relevant technology in the preferred formulation.

Since replacements for TiO_2 to achieve standard opacity at lower cost have been the prime consideration, any change in spreading rate achieved at constant dry hiding should be taken into account. A good basis for comparison of relative effectiveness of alternative technologies is to establish raw material cost per square metre at a constant contrast ratio.

5.3 Relevance of packing considerations

In examining these various alternative technologies, consideration of the way in which the key ingredients pack within the dry film will assist correct application to a particular formula. The relative volumes and particle sizes of the various ingredients, which will influence the way in which packing can take place, should also be taken into account.

Typical particle sizes of key ingredients are shown in Table 5.2.

Table 5.2 Particle size of key ingredients

	Diameter (μm)
Titanium dioxide	0.2
Extenders	0.05–20.0
Latex binders	0.1–2.0
Pittment microvoids	0.5–1.5
Spindrift beads	< 5.0 or < 25.0
Plastic pigment	0.2–0.5
Opaque polymer	0.4 or 0.55

Using the volume ratios given in Figure 5.1, the relationships of key ingredients at three different PVCs, 20%, 35% and 55% are shown in Table 5.3. From the following assumption of mean particle diameters, individual particle volumes can be calculated and are given in Table 5.4

Table 5.3 Relationship of key ingredients

	Volume % of PVC in dry film		
	20	35	55
Titanium dioxide	20	25	20
Fine particle extender	—	10	10
Coarse extender	—	—	25
Latex solids	80	65	45
	100	100	100

Table 5.4 Individual particle volume

	Diameter (d)	Radius (r)	r^3	Volume ($= r^3 \times 4.19$)
Titanium dioxide	0.2	0.1	0.001	0.00419
Fine particle extenders	0.8	0.4	0.064	0.268
Coarse extender	5.0	2.5	15.625	65.469
Latex solids	0.5	0.25	0.0156	0.0654

Dividing the volumes of individual materials in the selected formulations by particulate volumes gives a measure of the relative numbers of particles in a unit cube of each material, as shown in Table 5.5.

Table 5.5 Relative numbers of particles

	Volume % of PVC		
	20	35	55
Titanium dioxide	4773	5967	4773
Fine particle extender	—	37	37
Coarse extender	—	—	0.4
Latex solids	1223	994	688

Taking the simplest case of the paint at 20% PVC, the packing in a unit cube of 100 μm^3 can be visualised by considering one layer in that cube, which will have a side of 4.65 μm, as shown in Figure 5.3. The latex particles are assumed to pack in a rhombohedral network, giving approximately 110 particles per layer with 11 layers in the cube. In an idealised situation, the smaller diameter

Figure 5.3 Packing at 20% PVC.

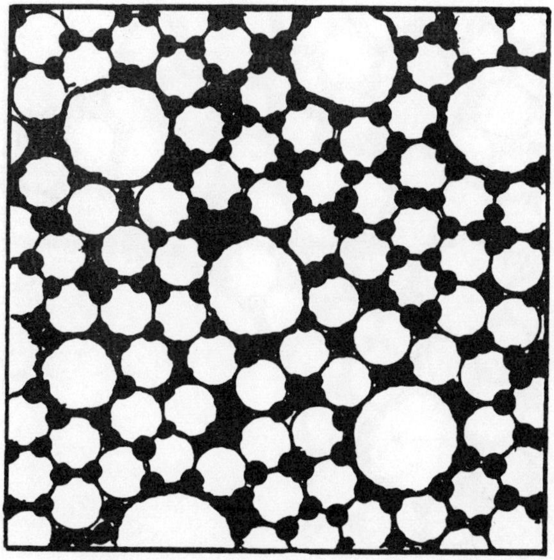

Figure 5.4 Packing at 35% PVC.

titanium dioxide particles will be uniformly dispersed at the interstices between latex particles, requiring approximately 2500 particles to give single spacing without flocculation or aggregation. In practice better than duplets is rarely achieved, so the 20% PVC paint may be considered as being as close to an optimum packing arrangement as is likely to be achieved with the titanium dioxide particles separated fairly uniformly by approximately their own diameter. This represents the situation in a latex gloss paint, which tends to approach as closely as possible the optimum achievable titanium dioxide utilisation. In such formulations, maximising gloss results from minimum flocculation.

Moving on to the semigloss formulation at 35% PVC, the 37 particles of fine extender can be thought of as replacing 148 of the latex particles. Within one layer of the cube 9–10 extender particles may be seen, well separated by a continuous matrix of latex particles with contained TiO_2 particles. However, in this case there are 25% more titanium dioxide particles occupying a volume of matrix 10% less than in the previous case. So while gross flocculation or aggregation is not expected, the TiO_2 is not being utilised as effectively as possible and there is scope for improvement.

In the case of the paint at 55% PVC 25% of the unit cube is occupied by 40% of a single large extender particle. Such particles create what is often referred to as the 'window effect', i.e. a significant volume of the film in which there is no possibility of light scattering and allowing incident light to approach the substrate without diffraction. Of the remaining space, 10% is occupied by fine extender particles, leaving only 65% of the total volume in which to disperse the same quantity of TiO_2 as in the first example. There is a clear opportunity for improvement in formulation here.

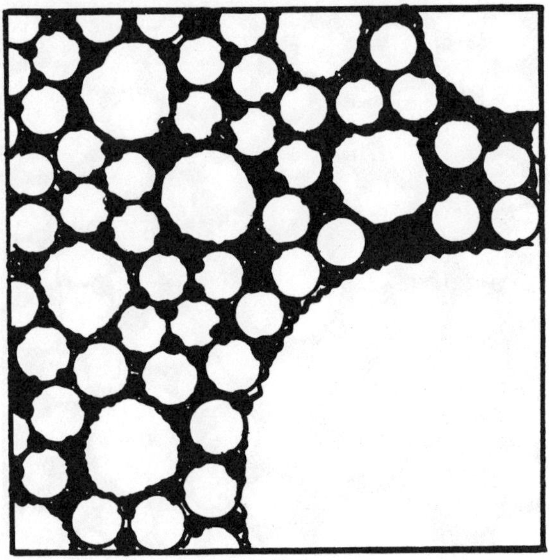

Figure 5.5 Packing at 55% PVC.

It should be emphasised that these examples and calculations are used to illustrate the principle of visualising spatial relationships within the film and particularly to recognise the significant impact of extender particle size on film packing. In practice a wide particle-size distribution, flocculation or aggregation will also affect packing, particularly in the case of many extenders, and combinations of extenders of different shapes may be used which will further influence packing arrangements.

The alternative technologies considered as means of TiO_2 reduction may now be better understood.

5.4 Pittment

Pittment is the trade mark of PPG Industries applied to their patented process for the incorporation of closed cell microvoids into coatings. Three possible routes were considered:

1. Leaching.
2. Incorporation of droplets of non-solvent or encapsulated air bubbles, very often by an emulsification technique.
3. Formulation of a gel structure containing dispersed non-solvents.

Of these the second route was preferred for latex paints. An emulsion of a volatile liquid which is a non-solvent for the polymer is formed in a polymer latex. This mixture, on drying, gives a continuous coalesced film of the polymer which contains dispersed droplets of the volatile liquid. Subsequent complete drying of the film results in discrete isolated microvoids. It follows from this brief description of the concept that there are a number of constraints. The volatile liquid used as a basis of the production of microvoids must be not only a non-solvent for the film former, but also be immiscible with water. It must have an evaporation rate slower than water, but should be capable of volatilising from the film under normal ambient conditions within 24 hours. In practice, it was found that the optimum particle size for the volatile liquid dispersion, and hence the microvoids, was in the range of 0.5–1.5 μm. Since these microvoids must maintain their identity once the volatile liquid has evaporated, it follows that the matrix of the paint must be rich in latex particles, i.e. must be formulated below the CPVC. It is found, however, that the presence of the microvoids tends to reduce the sheen and it is not possible with this technology to formulate high sheen or semigloss finishes. This limits the technology to low sheen and well bound flat paints in the 35–60% PVC range, depending on the choice of latex binder and other aspects of the pigmentation. Claimed advantages of the technology are improved mechanical properties with increased film flexibility, improved weatherability, less chalking and increased efficiency of light scattering.

In formulating paints on this technology, the microvoids can be regarded as fine particle extenders, but with virtually zero binder requirement. Coarse extenders should be excluded and the remaining pigmentation reduced to compensate for the presence of the microvoids. Since the presence of the microvoids increases the matrix volume significantly, it is possible to reduce the percentage solids volume whilst maintaining dry film build and opacity. In practice, up to 32% of the total dry film volume may be replaced with microvoids, which effectively increase the latex porosity, contributing to dry hide, whilst maintaining essentially a well bound film with low film porosity.

A typical formulation is given in Figure 5.6 which also illustrates the likely packing and spatial relationships in such coatings.

It has been noted that Pittment coatings have a brightness or purity of colour not normally experienced with films containing conventional extenders. Possible deficiencies are a tendency to give sheerness in the dry film, if applied by brush, and some reduction in ease of application. The large amount of non-solvent in the coating contributes an odour problem, particularly during the drying stages, which suggests a greater potential for the technology in exterior coatings, particularly in view of the claimed advantages in film properties. Finally, the potential problem of disposal of washings cannot be ignored.

5.5 Spindrift

Sprindrift was originally a development of Dulux Australia Ltd., but subsequently licensed to Tioxide International for commercial exploitation worldwide. The basis of Spindrift is a system of dispersed polymeric beads, gener-

Dispersed non–solvent droplets 0.5 — 1.5 micron

Formulating range 35 — 60% PVC

Typical formulation *Key*

TiO_2	16% —	•
Extender	18% —	
Binder	34% —	
'Non–solvent'	32% —	

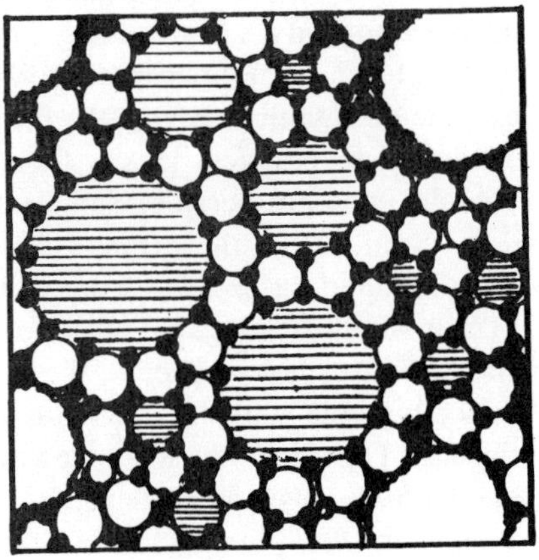

Figure 5.6 Pittment, typical formulation.

ally uniform in size and with a maximum diameter of either 5 or 25 μm, depending on grade. Typically the beads are prepared from a copolymer of styrene and an unsaturated polyester. The beads contain sealed pores or 'vesicles', which compromise a large proportion of the total bead volume. The vesicles may vary in size considerably, but a mean diameter of 0.6 μm is typical. The vesicles contain air and may also contain particles of rutile titanium dioxide pigment, which may also be present in the resinous portion of the bead. It is claimed that the controlled concentration of voids and dispersed titanium dioxide pigment within the beads gives high opacity and improved utilisation of the titanium dioxide employed. In practice, the commercial product supplied as an aqueous dispersion contains some titanium dioxide pigment external to the bead within the aqueous phase of the slurry.

In formulation, with Spindrift as with Pittment, it is customary to replace all coarse extender and titanium dioxide by the slurry of Spindrift, using fine extender to adjust the overall PVC to the required level. Typical formulations contain 30% by volume of Spindrift beads and between 25% and 50% of fine extender, giving PVCs in the range of 55–80%. At the higher PVCs it is recommended that a high binding latex such as a styrene–acrylate copolymer should be used. In using these high formulating PVCs it should be remembered, as with other polymeric extenders, that the volumetric oil absorption is likely to be close to 0.4, giving a CPVC of approximately 70% compared with 40% for conventional fine particle extenders. This allows the use of higher PVCs while retaining good film characteristics. A typical 65% PVC formulation and visual indication of packing characteristics is given in Figure 5.7.

Vesiculated polymeric beads	5 micron or 25 micron
Typical pore size	0.6 micron
TiO_2 content : internal	8%
external	24%
Formulating range	55–80% PVC

Typical formulation *Key*

Extender	35%	—
Beads	23%	—
TiO_2 (eternal)	7%	—
Binder	35%	—

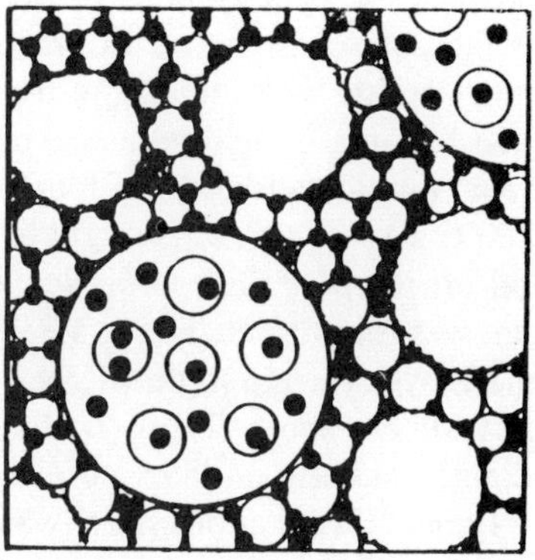

Figure 5.7 Spindrift, typical formulation.

The technology is limited to matt paints, mainly for interior use, since the presence of Spindrift beads tends to reduce longer term exterior durability. A benefit of the technology in interior paints is improved stain resistance and good burnish resistance. Use of the larger 25 μm grade gives very low angular sheen, which is regarded as a positive benefit in some markets. Further advantages of the technology are a contribution to good brushing properties and a potential benefit of a 'stir-in' pigment with no dispersion stage necessary, providing the selected fine extender can also be obtained in slurry form.

The main disadvantage of Spindrift is a slight reduction in colour and increased odour due to traces of residual styrene monomer. As with Pittment, the tendency is to use these technologies to gain product benefits at no increased cost, rather than to see them as a route to lower cost formulation.

5.6 Plastic pigment

The plastic pigment concept has been developed particularly by the Glidden Coatings and Resins Division of the SCM Corporation, now part of the ICI International Paint Group. Triggered by the original conference paper and subsequent publication of Burrell's work, a number of possible routes were investigated, but at an early stage the plastic pigment concept was showing such promise that all other routes were abandoned. Essentially this is a dispersion of solid sub-micron, non-film forming, polystyrene particles prepared by emulsion polymerisation, which are used to replace a percentage of the titanium dioxide and conventional extender in silk or matt finishes.

Although in the early work it was shown that the particle size of the non-film forming particles could be varied from 1550–7400 Å (0.15–0.74 μm) without significant effect on binder requirements, a balance had to be accepted between the opacity benefit of the finer particles and the film flow benefits of the coarser particles. In practice formulation was made on particles very much in line with the size of the main binder, i.e. around 0.4–0.5 μm.

In formulating paints using plastic pigment as a replacement for conventional pigmentation it was found that an improvement in opacity occurred as CPVC was exceeded, as might be expected, but that there was not a coincident fall-off in other film properties. Film porosity, as measured by enamel hold-out, was found to change at a PVC significantly higher than the CPVC. Ramig and Floyd referred to this higher critical PVC as the 'critical surface porosity' or CSP for short. These findings are in accord with Stieg's work on the distinction between latex porosity and film porosity (see Chapter 4).

Typical formulations for silk and matt paints using plastic pigment are given in Figure 5.8 which shows once again that approximately 30% of the total solids volume can be replaced by the alternative technology, with benefits in cost and in obtaining films with good stain resistance and cleanability.

Non film forming polymeric beads *c*.0.4 micron

Formulating range 30–85% PVC

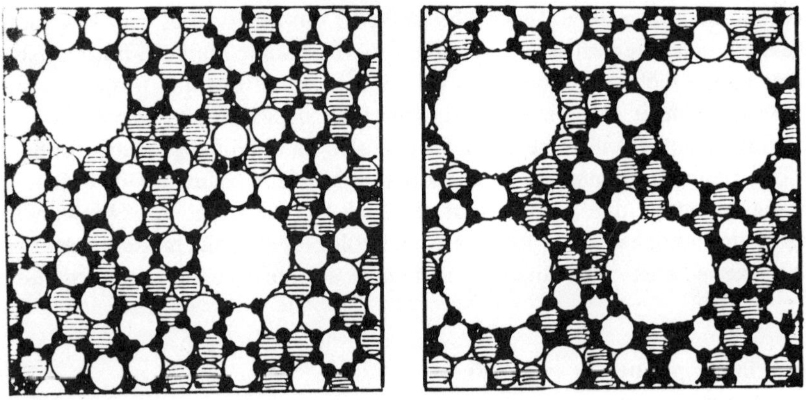

Typical formulation

	Silk	*Matt*
TiO_2	20%	20%
Extender	5%	20%
Binder	45%	30%
Polymerbead	30%	30%

Key

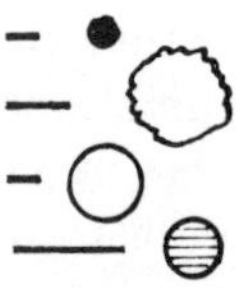

Figure 5.8 Plastic pigment, typical formulation.

5.7 Opaque polymer

Opaque polymer (now known as Rhopaque OP62) was developed by Rohm and Haas as the optimum in polymeric extenders, with a positive contribution to opacity without sacrifice of film properties. Essentially, Rhopaque is a non-film-forming emulsion polymer with a hollow core approximately 0.3 μm in diameter. In its earliest version the average external diameter was 0.55 μm, giving a wall thickness of 0.12 μm. Later developments allowed this wall thickness to be reduced to approximately 0.05 μm, giving a typical hollow sphere 0.4 μm in diameter. The acrylic–styrene beads are supplied in emulsion form, at which stage the beads are filled with water. When a paint containing Rhopaque OP62 dries, the water permanently diffuses from the core of the particles and is replaced by air, which then acts as a fully bound scattering site. The advantages claimed for opaque polymer are:

- A direct contribution to hiding.
- The ability to formulate paints with equal resistance properties at higher PVC, because of its lower binder requirement.
- Improvement in the scattering efficiency of the titanium dioxide by improved spacing of the TiO_2 particles.

As with the other technologies discussed, it is important to reconsider the total formulation if use of the new technology is to be optimised. Attempts to 'slot in' the new material to an existing formulation are unlikely to produce any positive benefits and could lead to the rejection of a potentially valuable formulating tool. Rhopaque OP62 was slow in gaining acceptance, as a result of this need to think in new ways, but its overall benefits, particularly in silk finishes, have now been extended to flat finishes and the product is gaining in acceptance worldwide.

Various guidelines have been developed to assist in the optimisation of opaque polymer-containing formulations, ranging from sophisticated computer programs to the use of matrices or 'reformulating protocols'. The computer approach will be referred to in Chapter 6. Within this chapter the general protocol now established as the simplest route to achieving a balanced formulation is considered. It must be recognised, however, that in different world markets these guidelines may be modified slightly to compensate for the differences existing in relative prices of key raw materials. For example, in the UK market at the time of writing, it is generally recognised that the maximum level of opaque polymer to incorporate into a formulation is 15% of the total solids volume. In the USA, the current recommendation is to formulate up to 20% PVC for opaque polymer. The limiting level will be influenced particularly by the relative prices of OP62, the binder used and the market price for titanium dioxide. If cost were not a limiting factor, it could be argued that the most efficient use of opaque polymer would be achieved at 30% PVC.

Hollow polymeric beads	*c.* 0.6 micron
Internal voids	*c.* 0.4 micron
Formulating range	30 – 90% PVC

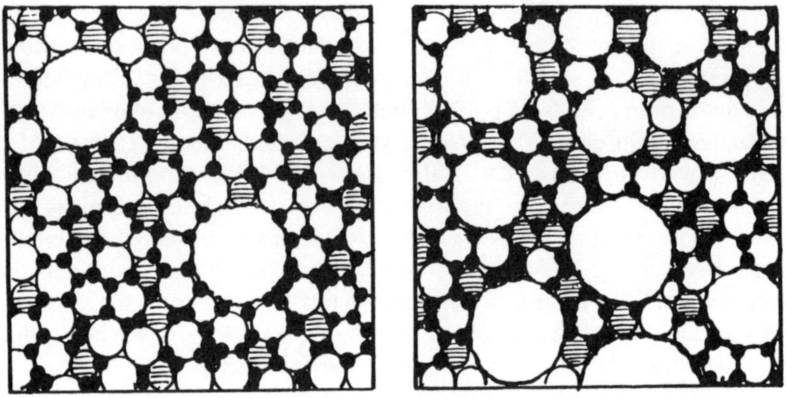

	Silk	*Matt*
TiO_2	20%	15%
Extender	5%	35%
Binder	60%	35%
OP	15%	15%

Key

Figure 5.9 Opaque polymer, typical formulation.

To return to the optimisation of formulations with opaque polymer, it must be remembered that it is a small particle size, non-crowding extender, it has the ability to scatter light and act as a pigment, it has low binder demand which effectively increases the CPVC of the formulation and it has an inherent benefit in scrub resistance. These features allow for reformulating a paint at reduced TiO_2 level and higher PVC whilst maintaining hiding and general resistance properties. Experience has shown that 4 PVC % of Rhopaque OP62 is approximately equal to 1 PVC % of titanium dioxide in contribution to hiding in most formulations. This leads to the following protocol:

1. TiO_2 PVC. If greater than 22%, reduce by 25%; if less than 22% – reduce by 20%.
2. OP62 PVC. If formula PVC greater than 55%, add five times amount of TiO_2 removed, subject to a maximum of 15% PVC or minimum of 8% PVC. If formula PVC less than 55%, add four times amount of TiO_2 removed, subject to a maximum of 20% PVC.
3. Functional extenders. These are defined as extenders, such as mica or talc, present for specific reasons, such as rheology control, durability enhancement, etc. If less than 10% PVC, keep constant; if greater than 10% PVC, reduce to 10% PVC.

4. Ultrafine extenders. If greater than 10% PVC, reduce by one third; if less than 10% PVC, reduce by one half.
5. Non-functional extenders. Adjust level to achieve required PVC at original total volume solids.
6. Binder latex. Adjust to required PVC.
7. Dispersant level. Adjust to a constant level on total inorganic pigments and extenders, *not* including opaque polymer.
8. Coalescing solvent. Adjust to a constant percentage on polymer solids, including binder and opaque polymer.
9. PVC. Initially adjust at 3% higher than original formulation, calculating OP62 as pigment. It may be possible to increase PVC still further to +5% or +8% over original formulation. To assess optimum level compare the tint strength of the modified formulation with the original formulation and select PVC which gives equal tint strength and hiding.
10. Grade of TiO_2. In existing formulations using lower levels (3–7% PVC) of heavily coated grades of TiO_2 (typically 80% TiO_2 content), it may be advantageous to replace with a standard enamel grade on an equal weight basis. This will assist in maintaining wet hiding properties.

Typical formulations for matt and silk finishes, using opaque polymer, are shown in Figure 5.9, again indicating the likely packing characteristics.

Bibliography

Pittment

V.I. Fartunin, G.V. Novoselskaya, N.J. Gribkova and S.V. Yakubovich, Preparation of coatings with microvoids and examination of their properties, Paper, XIIIth FATIPEC Congress, Cannes (1976) 206–9

PPG Industries, Production of void containing polymer films using emulsified, high boiling non solvent. US Pat. 3,915,726; *Off Gaz.* **939** (4), (1975) 2056.

PPG Industries, Opaque non pigmented microporous film and process and composition for preparing same. US Pat. 4,007,141; *Off Gaz.* **955** (2), (1977) 589.

M.S. Ramaiah and W. Funke, Thermal stability of microvoid coatings, *JOCCA* **64** (5), (1981) 191–7.

J.A. Seiner, Microvoid coatings: material and energy savers? *JOCCA* **60** (9), (1977) 335–47.

Spindrift

Dulux Australia Ltd. Matt paints (containing vesiculated polymer granules), Brit. Pat. 1,395,065. Addn to 1,288,583.

K. Goldsbrough, L.A.Simpson, and D.F. Tunstall, Formulating paints with pigmented vesiculated beads, *Progr. Org. Coatings*, **10** (1), (1982) 35–53

R.W. Hislop and P.L. McGinley, Microvoid coatings: pigmented vesiculated beads in flat latex paints, *J Coatings Tech.* **50**, 642 (1978) 69–77.

Tioxide Group Plc, Vesiculated polymer beads, Europ 081,306, 17pp. *Europ. Pat. Rept.* **K**, (25), (1983) Gp. G, 2.

Tioxide Group Plc, Cross linked polyester resin vesiculated beads. Europ 083,188, 16pp. *Europ. Pat. Rept.* **K**, (28), (1983) Gp. G 4.

Plastic pigment

Berger Jenson and Nicholson, Polymer aggregates as pigments and extenders—Ger 2,747,665: *Paint Resin Pat.* **15**, (8), (1978), Abs 790.

F.L. Floyd and A. Ramig, Plastic pigment: a novel approach to microvoid hiding II. Controlling parameters for performance, *J. Coatings Tech.* **51** (658) (1979), 75–85
F.L. Floyd, *Plastic pigments III. Influence of extenders and coalescents,* ACS Div. of ORPL, Papers **43** (1980) 31–8.
NL Industries, Opacifying polymeric particles. Brit. (A)2,106,915, 8pp.
A. Ramig and F.L. Floyd, Plastic pigment: novel approach to microvoid hiding. I. Introduction and mechanism. *J. Coatings Tech.* **51**, (658), (1979) 63–74.
A. Ramig and P.F. Ramig Plastic pigment: novel approach to microvoid hiding; effect of composition on latex paint performance. *JOCCA*, **64** (11), (1981), 439–47.
A. Renner, Stable, colloidal condensation polymers from urea and formaldehyde as new white pigments in emulsion paints, *Coating* **12**, 10,254 (1979), (3pp) (in German).
SCM Corp., Latex paint containing plastic pigment. Brit pat. 1,516,802 (29pp).
SCM Corp., Polymer particles as latex paint pigments. Dutch 76/04,396: *Paint Res. Pat.* **15**, (1), (1978), Abs 14
F.B. Stieg, Opaque white pigments in coatings, *ACS Symposium Series 285*, (1985) 1249–69

Opaque polymer

R.W. Andrew and B. Lestarquit, New additive helps cut the cost of hiding, *Polym. Paint Col. J.* **174**, (4122), 442 (1984) (3pp.)
C.A. Cant and R.G. Aviles, Quality semigloss paints at a very low cost, *Resin Rev.* 1986, **36**, (2) (1986) 11–8.
D. Dowling, B. Grange and B. Lestarquit, Opacifying polymers – a new raw material for the coatings industry, *Proc. XVIIIth FATIPEC Congress*, Venice 1986, Vol. I/A, 117–44
R.J. Equi, D.M Fasano and W.H. Hill, Use of opaque polymers in latex paints. *Am. Paint J.* **72**, (43), (1988) 85–8.
D.M. Fasano, J.W. Hook, W.H. Hill and R.J. Equi, Formulating high PVC paints with opaque polymer additives, *Resin Rev.* **37**, (2), (1987) 21–31.
D.M. Fasano. Use of small polymeric microvoids in formulating high PVC paints. *J. Coatings Tech.* **59**, (752), (1987) 109–16.
J.J. Gambino, W.J. Gozdan and P.M. Finegan, Meeting all expectations: Ropaque OP42 and Rhoplex AC829, *Resin Rev.* **35**, (1), (1985), 7–18.
W.J. Gozdon and H. Kroeker, Exterior durability of a properietary opaque polymer, *Farg Lack* **35**, (6), 113, (1989) (3pp).
B. Grange, S. Riemann and A. Rademacher, Opaque polymers in paints with high pigment volume concentration. *Farbe Lack* **93**, (10), (1987) 808–12.
A.R. Hatter and K. Nyi, New additive helps cut the cost of hiding, *Res. Review* **33** (2), (1983) 3–8.
A.R. Hatter, K. Nyi and P.M. Finegan, Formulating with an opaque polymer, *Res. Review* **33**, (2), (1983) 9–16.
W.H. Hill, D.M. Fasano and P.J. Rogers-Moses, A more efficient additive for lower cost hiding, *Resin Rev.* **35**, (3) (1985) 3–10.
J.W. Hook and R.E. Harren. Opaque polymers. *IX International Conference in Organic Coatings, Science & Technology.*, Athens (1983), 65–77
J.W. Hook, M.B. Freedman and J.T. Mooney, Wet hiding in paints containing opaque polymer modifiers. *Resin Rev.* **36**, (3), (1986) 3–7.
D.J. Kerrison Improved profitability through recent polymer innovation., *Polym. Paint Col. J.* **173**, (4093), (1983), 255.
F. Marschall and D. Marschall, Urethane associative thickener plus non pigmentary hiding polymer equals timely additives. *Am. Paint J.* **72**, (11), 76 (1987) (14 pp).
D.G. Milne, Non pigmentary hiding additive, *Am. Paint J.* **66**, (43), (1982) 46–9.
K. Nyi, Opaque polymer: light scattering, hiding additive for latex paints, *Austral OCCA Proc. News* **19**, (11), 4 (1982), (8 pp).
Rohm & Haas Co., Polymeric pigment particles, Europ 022,033, *Europ Pat. Rept.* No. 55, (1981) 554.

Extender

D.O. Adrien, Uniqueness in formulating, dedication, *Am. Paint J. Conv. Daily* **72**, (14), (1987) 101.

S.Y. Chang and W.E. Scott, Investigation into the mechanism of titanium dioxide extension by calcium carbonate, *Tappi* **71**, (10), (1988), 128–31.

M. Cremer, Results obtained with blanc fixe micro in high quality paints. *I. Polym Paint Col. J.* **176**, (4165), 310 (1986) (5pp); *III. Polym. Paint Col. J.* **176**, (4173), (1986) 654 (3 pp).

L. Cutrone, Influence of fine particle extenders on optical properties of latex paints. *J. Coatings Tech.* **58**, (736), (1986) 83–8.

K.A. Haagenson, Effect of extender particle size on the hiding properties of an interior flat latex paint, *Am. Paint J.* **72**, (43) (1988) 89–94.

D.L. Light and D.G. Jeffs, Investigation into effects of particle size on optical properties of paint films containing flash calcined clay. *XIVth FATIPEC Congress*, Budapest 1978, 397–403.

B. Martensson, Formulating with aluminium silicate spacer pigments, *Pig. Resin Tech.* **14**, (11) (1985), 4–6.

K.V.S.N. Raju and M. Yaseen, Effects of part replacement of titanium dioxide by barytes on the properties of alkyd paints, *JOCCA* **65**, (11) (1982) 401–8.

T.E. Remmers, Extend titanium dioxide with synthetic calcium silicates, *Am. Paint J.* **72**, (43), 83 (1988) (3 pp).

A. Rutherford, White carbonate extender in surface coating applications, *Paint and Resin* **58**, (2), (1988) 13–14.

R.F. Storey and J.E. Marcellus, Alumina trihydrate trims titanium dioxide costs in flat latex paints, *Mod. Paint Coatings* **78**, (2), (1988) 32–4.

Formulation

J.R. Chalmers and R.J. Woodbridge., Air and polymer-extended paints, *Polym. Paint Col. J.* **173**, (4104), Europ. Suppl. 94 (1983) (6 pp).

R.W. Craft, Calculation approach to efficient titanium dioxide formulation in coatings, *Proc. Paint RA 8th Internat. Conf.*, Optimising Paint Formulation, organised in conjunction with TNO Verfinstituut, Amsterdam (1988) 13–48.

M.S. El-Aasser, Microvoid coatings *Polym. News* **3**, (4), (1977), 209–14; *RAPRA Abs* **14**, (26), (1977) Abs 7705567L.

J.E Hall and V.R. Pedersen, One coat hiding: in theory and practice, *Surface Coatings Austral.* **25** (4) 12 (1988) (7 pp).

C.R. Hegedus and A.T. Eng, Coating formulation development using critical pigment volume concentration prediction and statistical design, *J. Coatings Technology* 1988, **60**, (767), (1988) 77–87.

D.A. Holtzen and W.H. Morrison, (jr) Statistical model for emulsion paint above the critical PVC, *J. Coatings Tech.* **57**, (729), (1985) 37–45.

D.A. Holtzen and W.H. Morrison, (jr), Statistical model for emulsion paint above the critical pigment volume concentration, *Proc. Paint RA 8th Internat. Conf.*, 'Optimising Paint Formulation', organised in conjunction with TNO Verfinstituut, Amsterdam (1988) 48–56.

W.D. Meadows, Critical pigment volume concentration (CPVC) of latex paints and formula optimisation II, *Am. Paint J.* **70**, (5), (1985) 42–8.

M.S. Ramaiah and R. Vittal Rao, Microvoids in organic coatings, *Progr. Org. Coat.* **11**, (3), (1983) 267–85.

R.H. Rowland and F.B. Stieg, Graphical solution to CPVC problems in latex paints, *J. Coatings Tech.* **54**, (686), (1982) 51–6.

J.A. Seiner, Microvoids as pigments: review, *Ind. Eng. Chem., Product R&D*, **17**, (4), (1978) 302–17.

F.B. Stieg, The ABC'S of white hiding power, *J. Coatings Tech.* **69**, (630), (1977), 54–8.

F.B. Stieg, Numerical expression of film porosity, *Am. Paint J.* **63**, (13) (1978).

F.B. Stieg, Avoiding excessive titanium costs, *J. Coatings Tech.* **53**, (682) (1981).

D.G. Tsimboukis, The packing of spherical particles analytical approximations. *6th International Conference in Organic Coatings Science and Technology*, Athens. (1980) 781–99.

Other

W. Funke, Possible uses of reactive microgels in paint formulation, *JOCCA*, **60** (11), (1977) 438–45.

M. Kerker and D.D. Cooke, Pigmented microvoid coatings. II Luminance and colour of the concentric sphere model, *J. Coatings Tech.* **48** (621), (1976) 35–41.

6 Computer techniques

LUIGI CUTRONE

6.1 Introduction

Micro- and minicomputers, introduced in the late 1970s, have come a long way from the enormous Eniac, the first computer in history, which was as big as a house, had a 30 metre high panel, 18 000 vacuum tubes and an operating capacity of 38 nine-digit divisions per second. In comparison, today's supercomputers can carry out billions of operations per second and be carried around in a briefcase. These small units, with enormous computing power, are now affordable by large and small organisations as well as individuals. The availability of easy, user-friendly and inexpensive software has contributed to the increased usage of computers. In the past people were reluctant to use them; now computers are an integral part of daily activities. The acceptance of computers is partly due to the well designed and user-oriented software, and to the good marketing strategy of the computer industry. Most of the computer applications in the coating industry evolved by adopting and adapting existing software such as data bases and spread sheets developed for other sectors. Hence, it is not surprising that the first computer applications in the coating area were formulation, costing programs, batch tickets, inventory control and the performance of office procedures.

These findings were highlighted in a survey carried out in 1982 by the New York Society for Coatings Technology [1]. The study comprised a survey of 800 paint manufacturers from the USA, Canada and United Kingdom. Of that total, approximately 25% responded. The findings revealed that a significant number of paint companies used computers, and that 85% of those answering the questionnaire were satisfied with their computer equipment. More than half the respondents used their equipment for costing and formulations programmes.

Other important findings were:

- Nearly 80% of the respondents owned their own equipment, which consisted of mini and mainframe computers.
- One third of the respondents developed their own software, and another third purchased their software from system houses. The major computer language used, especially in small and medium size companies, was BASIC.

- The most common use in laboratories was to determine formulation parameters. Very little reference was made to development work and specialised process control.

In the last few years the use of computers in the coating field has increased at a great rate, especially in process control [2] and research and development [3], with more specific and tailor-made software.

Formulation programs include a series of software designed to help formula-based manufacturers create and utilise continuous communications channels between departments, as well as to establish and maintain control of operation processes. The software includes formulation management, which defines and maintains the master formulation file; costing and formulation parameters, which help the formulator in costing, rebulking, yield and other formulation constants; raw material management, which maintains a list of all the raw materials including minimum stock and reordering schedules and often also the generation of the material safety data sheet (MSDS) [4–6].

Another area of computer application has been in process control [7–9]. This, however, is not an easy task, because, among the many problems facing the system builder who sets out to create an automatic control system for high volume paint production are the complexities of creating, modifying, storing and implementing the hundreds of formulations, each with up to 20 liquid and solid ingredients. Despite the difficulties, there are many automated control systems which enable paint manufacturers to operate several mixers at the same time, producing many different formulations.

Much more progress has been achieved in the colour matching field where the existing software not only gives the proper constituents for matching a specific colour, but at minimum cost and the lowest degree of metamerism [10–12].

The most recent coating area where computers have been introduced with success is in research instruments [13] and formulation development. All the major instrument producers have now incorporated a microprocessor with the instrument so that the operator can obtain the analysed data quickly and in the desired format.

In product development, because of the lack of understanding of the relationship between paint components and performance properties and the inability to use theoretical models for formulation purely on a molecular basis, computer usage has been restricted. However, the computer has been used to generate empirical equations that correlate with experimental data through the use of experimental design [14]. Numerous computer-assisted techniques, particularly multivariate statistical procedures such as regression analysis, principal component analysis and discriminant analysis, help the formulator to generate equations which can be used to predict paint performance.

6.2 Computerised process control

Paints are produced from recipes by plants that have a series of parallel or in-line unit operations. Process automation will inevitably result in large savings from improvements in consistency, quality, throughput and elimination of human error. However, there are some practical difficulties to overcome. Most process computers are designed to perform continuous control algorithms such as volume and temperature, but paint manufacturing tends to be primarily sequential in nature. Some software producers offer packages developed for sequential control of process in paint manufacturing; however, to obtain satisfactory results a large amount of custom programming in addition to the configuration is still necessary.

A paint manufacturing process can be divided into three main areas: formulation management, task execution and batch reports [2]. Formulation management is the master file which contains all the information. Using simple languages, each formulation has to be entered into the data base, given an identification number and stored. The formulation includes the quantities of each ingredient and the plant actions required for mixing. Simplicity of programming is essential to allow users to make formulation changes quickly and easily.

The task execution file contains all the information pertaining to the specific task (i.e. charge, fill, mix, etc.). Some software tracks each batch through the plant to ensure that the formulation is executed correctly, alerting the operator to the process and drawing attention to any problems. If a batch sample is found to be out of tolerance it is possible to initiate an adjustment cycle to correct the process.

The batch report file, as the name implies, contains all the information about the product including physical and chemical analysis results, customer specifications and packaging.

Computerised process control can be complex to install, but the economics can be very favourable. There have been examples where production has doubled with improved quality due to the elimination of inconsistency from batch to batch arising from the manual control. This consequently increases throughput and reduces material and disposal costs.

6.3 Colour matching

Colour matching has traditionally been very subjective in nature and dependent entirely upon the eyes of trained colourists.

The human eye cannot separate the component colours of a light source and therefore it is possible to produce the sensation of any colour in an infinite number of ways. In 1853 Grassman [15] postulated that light of any colour can be matched by the combination of not more than three suitable primary colours, provided that none of them can be matched by any mixture of the other two. He further stated the principle of additivity: lights that give the same visual effect separately produce the same effect in mixture, even though the spectral compositions are different. Thus if two coloured lights can be matched by a1, b1 and c1, and by a2, b2 and c2 (parts of three primaries A, B and C) and the colour produced by mixing them has a3, b3 and c3 parts; then:

$$a3 = a1 + a2\,,\ b3 = b1 + b2 \text{ and } c3 = c1 + c2$$

To explore further the possibility of colour matching, these hypotheses were studied by later workers notably by Guild [16] and by Wright [17]. However, instrumental colour matching was only adopted in the 1960s due to the colorimetric and spectrophotometric techniques and the numerous calculations required [12]. Since then there have been some 100 colorimeters and spectrophotometers from around 20 manufacturers available on the European market [18]; and much colour matching software which can match any given colour for cost, metamerism and opacity.

6.4 Formulation and costing programs

To develop formulations, the relationship between the constituents of a mixture and the performance properties such as optical, rheological, mechanical and storage stability must be clearly understood. Some mixtures are extremely complex systems and there are no theoretical models that allow the formulator to approach the subject matter purely on the basis of molecular theory. Therefore, most of the software available for formulation is oriented towards the calculation of formulation constants and cost, and little software has been developed to assist the formulator in reformulating work involving physical and chemical properties.

However, this software has been very helpful to the formulator, because calculating raw material costs and formulating constants is a very tedious task. Changing even a single ingredient requires calculations of a long series of numbers to determine new values. Furthermore, formulating for a specific target property can take hours of trial and error, which may or may not produce the required product.

The initial work in this area was carried out by Williams and Bacchetta [19] who developed a computer program in FORTRAN IV. Later the Philadelphia Society for Paint Technology introduced a FORTRAN programme [20] of considerable complexity using IBM equipment. One of the first formulation program packages available commercially was offered by General Electric on time-sharing computers. The program was called PAINT$ [21].

Since then there have been a number of software packages to assist the formulator running on small pocket computers [22, 23] to micro- [24–28] and macrocomputers [29] networked with process and commercial departments. A number of these software programs are free and published in technical journals. Others, more complex, developed by paint specialist or software houses, are available at a cost depending on the software capability.

Independent of the type of hardware and software, formulation programs contain three main areas: raw material file, formulation file and calculation file.

6.4.1 *Raw material file*

For the computer to do the necessary calculations a comprehensive data base with all the pertinent information about the raw material is necessary. Therefore, this part of the software must enable the user to create a file with all the raw materials and the corresponding characteristics. Furthermore, the software must enable the user to edit the records and to list the information stored on particular records in any desirable format.

In order to speed up the execution of the programme, raw materials are assigned a code which consists of a letter from A to Z and a number from 1 to 9999. The letter indicates the type of material and the number is assigned progressively as the raw materials are entered.

The program assigns a class of raw material to each of the letters of the alphabet for the raw material codes. These classes are listed in Table 6.1. Once the files are created, the next step is to fill them. This is a simple process of answering a series of questions; a prompt about the material such as the name of the material, density, percentage solids, cost, etc.

Table 6.1 Raw material classification

A – Special additives	L – Latex resins
B – pH additives	M – Miscellaneous
C – Coalescents	N – Surfactants
D – Dispersants	O – Oils
E – Driers	P – Aliphatic solvents
F – Defoamers	Q – Aromatic solvents
G – Extenders	R – Rheology modifiers
H – Coloured pigments	S – Titanium dioxide (slurry)
I – Coloured pigments (suspension)	T – Titanium dioxide
J – Hard resins	W – Wetting agents
K – Solution resins	

With the raw material file, the software must offer the option of listing the information stored on any one particular raw material and the opportunity to correct the information stored, respectively. Correcting the information about a raw material must also be a simple task consisting of answering the minimum number of questions needed to correct the entry.

Lastly, there must be the option to list all the raw materials which have been stored, as shown in Table 6.2.

Table 6.2 List of raw material file

Code	Name	Specific gravity (wet)	Specific gravity (dry)	Solids (%)	Cost
A0001	Ad It	1.000	0.000	0.0	9.10
A0002	Acticide THP	1.136	0.000	0.0	1.72
A0003	H21	1.000	0.000	0.0	0.18
D0001	Orotan 731	1.104	1.420	25.0	0.95
D0002	KTPP	2.571	2.571	100.0	2.74
D0003	Dispex N40	1.300	1.750	40.0	0.90
G0001	Speswhite	2.600	2.600	100.0	0.11
G0002	Snowcal 6ML	2.700	2.700	100.0	0.10
G0003	Talc AT-1	2.900	2.900	100.0	0.34
K0001	P470	0.960	1.070	70.0	1.10
K0002	Epikote 1001	1.050	1.210	50.0	2.62
K0003	Desmophen 651	1.100	1.220	65.0	1.91
L0001	Viking 8020	1.070	1.135	55.0	0.45
L0002	Emultex VV565	1.070	1.120	52.5	0.45
L0003	Vinamul 3459	1.090	1.200	52.0	0.45
P0001	Isoparaffin E	0.723	0.000	0.0	0.85
P0002	Isoparaffin G	0.748	0.000	0.0	0.65
P0003	Isoparaffin H	0.759	0.000	0.0	0.72
T0001	Tioxide R-XL	3.700	3.700	100.0	1.50
T0002	Tioxide TR92	4.050	4.050	100.0	1.50
T0003	Tiona 535	4.000	4.000	100.0	1.40

6.4.2 *Formulation file*

Once the raw materials have been entered and coded, the formulation can be entered and the required calculation carried out. Ingredients are added one at a time by weight or volume; as they are entered the computer associates the name of the raw material with the code number in the raw material file. Once all the formulation ingredients have been entered and checked for any mistake, the information can be stored under formulation name and code or carry on formulation constants.

6.4.3 *Calculation file*

This file contains all the procedures for calculating the formulation constants. Here the information from the raw material file will be blended with the information from the formulation data base to generate the results of the calculations that are actually of interest. The information about the raw materials is taken and applied to the formulation in question. Here calculations of general interest such as specific gravity, pigment volume concentration, pigment-to-binder ratio, solids by weight, solids by volume, amount of TiO_2

	WEIGHT	VOLUME	SOLIDS WEIGHT	SOLIDS VOLUME	COST £
Pigment	206.0	50.9	206.0	50.9	0.309
Water	78.0	78.0	0.0	0.0	0.000
Foamaster NXZ	2.0	2.2	2.0	2.2	0.006
25% Orotan 731	8.5	7.7	2.1	1.5	0.008
Triton GR5	1.7	1.7	1.7	1.7	0.008
Propylene Glycol	100.0	96.8	0.0	0.0	0.500
2.5% Cellobond 5000	33.0	33.0	0.8	0.8	0.000
Nuodex 321E	0.8	0.8	0.8	0.8	0.004
0.88 Ammonia	1.0	1.1	0.0	0.0	0.001
Primal AC507	553.0	526.7	248.8	207.3	0.686
DALPAD A	16.0	14.5	0.0	0.0	0.023

Sp. Gr.	1.23
PVC	19.70
P/B	0.246
% Solids by weight	46.22
% Solids by volume	32.70
TiO2 per litre	253.26
RM cost per litre	1.90

Figure 6.1 Full gloss acrylic.

per litre and raw material cost per litre can be included; or more complex calculations, as seen later, such as replacement of raw materials, hiding power, flocculation, volatile organic compounds, etc. An example of the print-out of a formulation with the required formulation constants is shown in Figure 6.1.

The program discussed was originally on a time-sharing computer, and later adapted to a microcomputer using the BASIC language. Nowadays this type of task can be achieved using an electronic spread sheet with very little effort. This is only the basis for a formulation program, which can be extended in many different directions.

6.4.4 *Reformulation programs*

In the last few years more complex programs have evolved that take into account physical interactions of the components, for example hiding power, dry film porosity, packing, pigment extender and pigment thickener interactions [30–35].

Stieg [33] has developed a reformulating program running on an IBM computer or any IBM-compatible computer. The software allows the formulator to optimise existing formulations, on the basis of spacing theory (i.e. using spacing extenders) and minimise cost while maintaining the same level of quality.

Table 6.3 Structure of the raw material database required to run the program

1. For resin solutions or latex dispersions
   (a) total kg (wet-basis) per unit batch
   (b) density in kg/litre
   (c) cost in CU (currency units)/kg (wet-basis,
   (d) % solids by weight
   (e) latex binding power index (assume 0.65 if unknown)
   (f) solvent cost in CU/litre (solvent-base only)
   (g) solvent density in kg/litre
2. For titanium pigments
   (a) total kg per unit batch
   (b) density in kg/litre
   (c) cost in CU/kg
   (d) % titanium dioxide content
3. For all extenders (one SPACER, four others)
   (a) kg/unit batch
   (b) density in kg/litre
   (c) mean particle-diameter in microns
   (d) cost in CU/kg (wet-basis)
4. For opaque polymer (OP-62)
   (a) total kg (wet-basis)/unit batch
   (b) cost in CU/kg (wet-basis)
5. For additives (thickeners, surfactants, driers etc.)
   (a) total CU/unit batch
   (b) total kg/unit batch
   (c) total litres/unit batch
6. For overall formula
   (a) total unit batch size in litres

The software requires the creation of an extensive data base on all the paint additives as shown in Table 6.3. Once the data have been entered and the program has been loaded, the formulator then has all the reformulating options (Table 6.4).

The software allows the formulator a comparative evaluation on the basis of cost and performance characteristics for pigments of varying titanium content, extenders of different particle size and oil absorption, and for latex

Table 6.4 Reformulating options

A – Input of original	K – Relative tint strength
B – Re-pigment	L – Save program only
C – Minimise cost	M – Change porosity
D – Change % solids/volume	N – Recall original
E – Change hiding	O – Original dry hiding
F – Change % PVC	P – Change batch size
G – New formula	Q – Wet-to-dry HP change
H – Get E/d values	R – Repeat formula
I – Save program and data	S – Wet hiding power
J – Porosity match	T – Add opaque polymer

vehicles of different binding powers. Wet and dry hiding power can also be calculated even in the presence of dry hiding.

Analogous to the reformulating program, the Rohm & Haas Company [34] has recently developed and introduced a new program for the trade sales paint chemist for reformulating paints with organic hollow beads. This new software, called PC Formulate, contains mathematical models which allow the formulator to estimate the performance of paints containing hollow beads. The program is menu driven, using functions keys to carry out common commands, it is easy to edit and allows formulations to be stored.

Below critical pigment volume concentration (CPVC), the program contains mathematical models able to match the scattering power and estimate gloss of paints containing opaque polymer. This is achieved by searching for the best combination of titanium dioxide pigment, hollow beads and extender levels to give the required optical properties. An example is shown in Figures 6.2 and 6.3, where a high performance interior semi-gloss paint is reformulated with hollow beads keeping the same optical properties.

Above CPVC the software uses an empirical approach, which consists of the preparation of a few samples below and above the PVC following certain constraints and matching the performance properties of the control paint.

MATERIALS	PARTS BY WEIGHT	
Propylene Glycol	60.00	
Orotan 731 (25%)	9.40	
Ammonia (25%)	1.00	
Dowicil 75	1.00	
Water	30.00	
Polymekon 1488	3.00	PVC
TiO2 R-HD2	262.40	19.22 (TiO_2)
Durcal 2	100.50	11.04 (extender)
Primal HG-44	625.10	
Propylene Glycol	48.00	
Texanol	50.00	
Acrysol RM5/Water (1/1 pH 8.5)	70.00	
Ammonia (25%)	3.00	
Water	21.50	
	1284.90	

PVC	:	30.26
VOLUME SOLIDS	:	33.62
WET DENSITY	:	1.28

Figure 6.2 High performance interior semigloss emulsion paint.

	CONTROL PAINT ESG-44-2	OP-62 VERSION
KEY INGREDIENTS		
TiO2 R-HD2 PVC	19.22%	16%
DURCAL 2 PVC	11.04%	9%
OP-62 PVC	-	13%
FORMULATION CONSTANTS		
TOTAL PVC	30.26%	38.00%
VOLUME SOLIDS	33.62%	33.62%
WET DENSITY	1.28%	1.24%
PROPERTIES		
ICI—CONE AND PLATE	2.5	2.55
KU UNSH/SHEARED	128/112	121/116
CONTRAST RATIO (175 μm)	96.61	96.53
Y REFLECTANCE	17.10	17.40
SHEEN 60°/85°	35/61	36/67
% SAVINGS / LITRE	-	8.1%
TiO2 USAGE G/L	262.40	217.86
SAVINGS	-	17%

Figure 6.3 Reformulated interior semigloss emulsion paint.

6.5 Experimental designs

Paints, like many other products, are manufactured by mixing two or more ingredients. The properties and formulation costs of such products change significantly as the proportions of the ingredients are changed. The fundamental experimental problem in developing such formulations is the identification of those compositions which give the best balance between properties and formulation costs.

The most commonly used experimental strategy is to make up and evaluate a series of test mixtures of varying compositions and attempt to determine the optimum composition from evaluating the resulting data.

It is therefore imperative, before starting the experiment, to ensure that the series of samples to be prepared is properly set and that the experimental program is adequately designed to answer the questions as efficiently and precisely as possible. Unfortunately, what often happens is that the experimental program is inadequately prepared, resulting in wastage of effort and misleading calculations.

It is therefore useful to consider how this might be minimised in practice. With the introduction of computers, statistical techniques which existed but were not used because of the many calculations involved, have helped the investigator enormously in setting up an experimental design which yields maximum information with minimum effort and with great reliability. At the same time, they encourage detailed formulation of parameters and assist in eliminating sources of error arising from causes external to the experiment which were often beyond the control of the investigator. Proper experimental design involves analysis of the variance and ultimately in developing a model using appropriate equations [14].

In the classical approach to experimentation, it was common practice to vary one component, while holding the others constant, and study the effect of the variable component on the mixture. Such an approach is not only uneconomical, but might give rise to misleading results if the variables are dependent on one another. Therefore, a system which would enable more than one variable to be studied simultaneously is of great value to the investigator in saving time and determining any possible interdependence of the variables under study. This can be achieved by preparing a series of samples following simple rules, analysing the variance and postulating a model.

In preparing an experimental program two concepts are of great importance to the investigator. These are the interdependence of one variable on another, known as interaction, and error [36]. To illustrate the first consider two factors examined at two levels. For example, suppose additive A is examined at 5% and 10% and additive B at 1% and 2% respectively. The best results are obtained when additive A and additive B are used in the following concentrations:

A at 5% and B at 2%
A at 10% and B at 1%

whereas the worst results are when the additives are used at the levels of:

A at 5% and B at 1%
A at 10% and B at 2%

Following the classical approach, keeping the concentration of additive A at 10%, the best results are obtained by decreasing the concentration of additive B; whereas the opposite occurs if the concentration of additive A is at 5%. In this case the two additives are interacting with one another and the best results are obtained at an optimum level of A and an optimum level of B. This type of interaction is not infrequent in coatings and it is clearly important to establish the presence of such interactions to optimise the level of each constituent.

The other important factor to be considered in designing experiments is error. Errors in experimentation can arise in several ways and may be controllable or uncontrollable. Experimental errors can be broadly classified into two categories, systematic and random errors.

Systematic errors are those experimental errors whose numerical value tends to remain constant or follow a consistent pattern over a number of experimental runs. These may arise from something which remains constant throughout the experiment such as a single batch of raw material, a single operator, a single machine, or may arise from a consistent pattern or cycle such as hours within a day, season of the year, catalyst age and age of equipment.

Random errors are those experimental errors whose values change from one operator to another. Frequently, random errors follow approximately a bell-shaped Gaussian error distribution. The principal statistical tool for dealing with random errors is replication.

The distinction between the two types of errors can be seen from the following example. Suppose that a reflectance determination on a grey paint is carried out in duplicate on the same sample for six consecutive days as shown in Table 6.5. From the example it can be seen that all duplicates agree to 0.1 units of reflectance but the overall spread of results amounts to 0.6 units. The day-to-day error is much larger than that due to the method. It is also evident that data of this type enable estimates of two types of errors. This also demonstrates how experimental design might be set out to reveal the effect of different factors and also to eliminate unwanted effects.

Table 6.5 Reflectance measurement on grey paint samples

Day 1	64.0	63.9
Day 2	64.2	64.1
Day 3	63.7	63.6
Day 4	63.8	63.9
Day 5	64.3	64.2
Day 6	63.7	63.8

It is evident that the more determinations carried out, the more reliable is the mean of those results. It is equally true that when the error is large more determinations are required to achieve a reliable mean. Therefore, there are cases where duplication is not adequate and perhaps three or four determinations are necessary.

To minimise the number of samples and to identify and minimise errors several techniques of experimentation are available.

6.5.1 *Randomisation*

Randomisation ensures that determinations are made in a haphazard order decided purely on chance considerations and therefore the effect of random errors is taken into account. Randomisation is an extremely simple process to carry out. The most obvious way is to draw lots from a hat or use a pack of cards, or use tables of random numbers generated by computer. It is important, however, that each sample replica is considered a determination.

6.5.2 *Randomised block*

The simplest experimental design is the randomised block [37]. This involves the concepts of randomisation and replication discussed earlier. This type of experimentation might not be very useful to the coatings industry, but illustrates very useful concepts which are necessary to understand the more complex designs.

In setting up the experiment, suppose that the investigator had to compare four treatments (e.g. four temperatures in a polymerisation reaction or four plasticisers in a plastic material) and the treatments will be applied to a material which can be segmented into blocks, where four experiments will be carried out. If the treatments are labelled A, B, C and D and the blocks are 1, 2, 3 and 4, the first experimental design could be as shown in Table 6.6.

Table 6.6 Experimental design no.1

Block	Treatment
1	A A A A
2	B B B B
3	C C C C
4	D D D D

Little information can be obtained from the first experimental design because the differences observed are due either to the difference between the blocks or the difference between the treatments. A more refined experimental design would be to randomise the treatment as shown in Table 6.7.

Table 6.7 Experimental design no.2

Block	Treatment
1	D B B D
2	C D A B
3	B A C A
4	C C D A

Calculating the mean for each of the treatments will give a value that is statistically valid; however, the error in each mean will be inflated by the difference due to the blocks. The experimental design may be refined still further by allocating the treatment to each block and randomising each treatment within the blocks as shown in Table 6.8.

Table 6.8 Experimental design no.3

Block	Treatment
1	D B A C
2	B C A D
3	C B A D
4	A C D B

Now all treatments appear an equal number of times in each block and the replicates all appear in different blocks. This approach has shown that a series of experiments may be arranged to eliminate a source of error and to obtain an estimate of the reliability of the factors under study. It has, however, been pointed out earlier that with proper design it is possible to study more than one variable at a time. This can be achieved with the latin square.

6.5.3 *Latin square*

The latin square [37] is one of several experimental designs available for studying two variables simultaneously and is an extension of the randomised block design. The randomised block showed how to arrange four levels of one variable in such a way that the difference between blocks was eliminated. With latin squares the same treatment is applied to the vertical columns. An example of the design is shown in Table 6.9.

Table 6.9 Experimental design no.4

Block	1	2	3	4
1	A	B	C	D
2	C	D	A	B
3	B	A	D	C
4	D	C	A	B

The technique can be illustrated by the following example. Suppose that four millbases are let down in four different ways: A, B, C and D. The design would be as described in Table 6.9 where the rows represent the millbases and the columns represent the treatments. To obtain the same results without the above design, one would have had to prepare four samples per millbase and to replicate this four times, thus giving $4^3 = 64$ experiments. The latin square design uses only 16 experiments even though there are four replicates of each treatment, thus saving considerable time and effort.

6.5.4 *Factorial experiments*

The latin square demonstrated how more than one factor can be studied in a series of experiments. It is, however, an essential condition that the factors do not interact with one another. This is not often the case in the coating industry, but can be overcome by using factorial designs.

Factorial designs are commonly used because they permit the estimation of the effects of several factors simultaneously. They are able to estimate the main effects as well as the interactions. The design is accomplished by making experimental runs at all combinations of factors m with l levels per factor. The number of experiments required would be $n = l^m$. The number of experimental runs can become numerous if l or m are large. In general the most popular and useful is the two-level factorial design where the number of runs is given by $n = 2^m$. Even the two-level experimental design can become very large; for example, for a six-factor, two-level design the number of samples required is 64 ($2^6 = 64$). For a larger number of factors, as seen later, a fraction of the full design may be appropriate.

For a two- and three-level factor, a geometrical figure can be drawn to visualise the factor space. With more factors, the cube becomes a hypercube and can only be interpreted mathematically.

For a three-factor experimental design the factors can be visualised quite adequately by using a cube, where the three factors are X_1, X_2, X_3, which are used at two levels and are represented by the eight corner points of the cube in Figure 6.4. Each design point represents a set of conditions. For example, at the four corners of the left side of the cube X_1 is at its low value, whereas at the four points of the right side of the cube, X_1 is at its high value. Denoting the high and low value by + and – respectively the experimental conditions at each corner of the cube are shown in Table 6.10 and Figure 6.5. For example, the left front lower corner represents the condition where X_1, X_2 and X_3 are at their low values as denoted by [– – –].

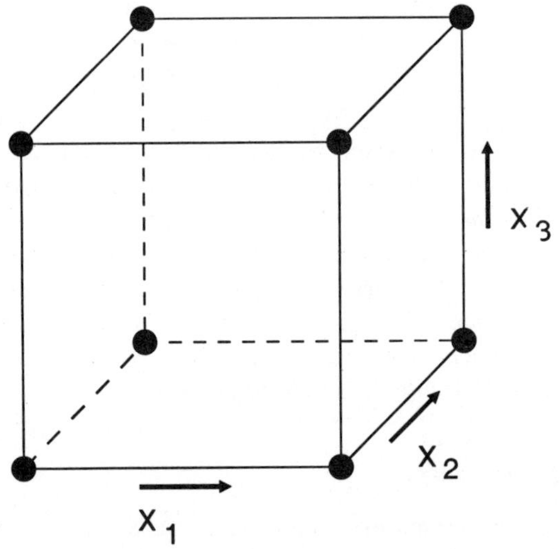

Figure 6.4 Two-level three-factor factorial design.

The technique can be best illustrated by going through an example. The objective is to determine the main effect and any interaction (if present) between titanium dioxide pigment, calcium carbonate extender and a rheology modifier on flocculation. In order to have an adequate number of degrees of freedom for error, a reasonable sensitivity for a test of overall curvature and an ability to detect with high probability any effects more than twice the experimental error, two replicates and four centre points are required, giving

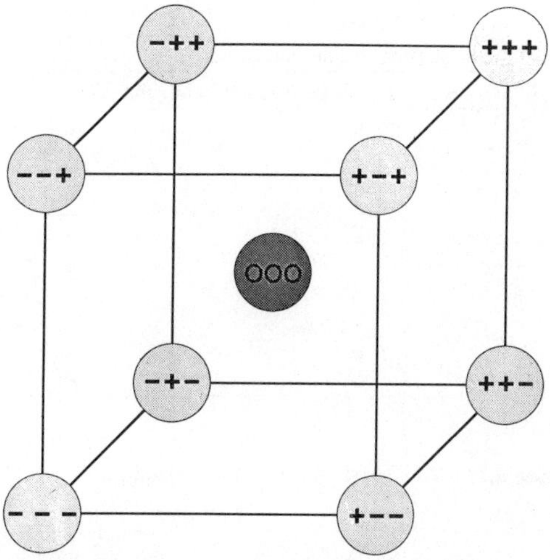

Figure 6.5 Two-level three-factor factorial diagram.

a total of 20 data points. From Table 6.11 the appropriate design (2^3) is selected and the random order of preparing the samples is taken from Table 6.12. A graphical representation of the experimental design is shown in Figure 6.6, where the corners of the cube represent the experimental conditions which are described in Table 6.13. The samples were prepared randomly using Table 6.12 as a guide.

Once the experimentation has been carried out, pigment flocculation is assessed using a method developed by Simpson and Rutherford [38, 39]. For

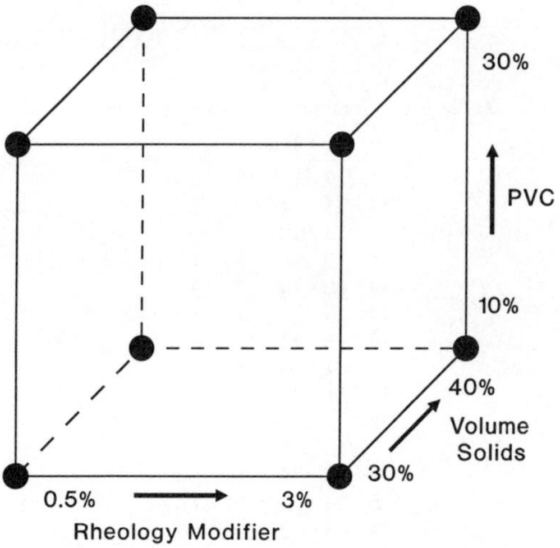

Figure 6.6 Factorial experiment to determine the influence of PVC, volume solids and rheology modifier on pigment flocculation.

Table 6.10 Experimental conditions

Trial	Experimental conditions for		
	X_1	X_2	X_3
1	−	−	−
2	+	−	−
3	−	+	−
4	+	+	−
5	−	−	+
6	+	−	+
7	−	+	+
8	+	+	+
9	0	0	0

Table 6.11 Two-level factorial design patterns

Trial	Factors			
	X_1	X_2	X_3	X_4
1	−	−	−	−
2	+	−	−	−
3	−	+	−	−
4	+	+	−	−
5	−	−	+	−
6	+	−	+	−
7	−	+	+	−
8	+	+	+	−
9	−	−	−	+
10	+	−	−	+
11	−	+	−	+
12	+	+	−	+
13	−	−	+	+
14	+	−	+	+
15	−	+	+	+
16	+	+	+	+

Table 6.12 Random orders for test trials

16 Trials				
3	9	11	10	3
1	15	1	7	5
11	16	12	11	10
5	12	4	5	1
8	10	14	8	14
4	5	3	4	16
12	13	15	15	8
15	1	10	14	13
10	11	13	3	9
2	4	6	2	11
7	6	8	13	12
9	8	7	1	6
6	14	5	9	7
14	3	2	12	4
16	7	9	16	2
13	2	16	6	15

each sample the value is recorded. Afterwards the mean (Y), the range and the variance (σ) are calculated for each observation and the standard deviation is calculated from the variance values. These data are necessary to determine the significance of the factor effects in the design. Due to the many calculations involved, analysis of the experimental data can be carried out using one of several computer assisted methods.

Table 6.13 Experimental conditions for the flocculation gradient determination

Trial	Variables (coded)			Variables (values)		
	X_1	X_2	X_3	X_1	X_2	X_3
1	–	–	–	0.5	30	10
2	+	–	–	3.0	30	10
3	–	+	–	0.5	40	10
4	+	+	–	3.0	40	10
5	–	–	+	0.5	30	30
6	+	–	+	3.0	30	30
7	–	+	+	0.5	40	30
8	+	+	+	3.0	40	30
9	0	0	0	1.5	35	20

The main effect can be calculated graphically by comparing the response values at the left and right planes of the cube as shown in Figure 6.7 for X_1, front plane with back plane for X_2 and bottom plane with top plane for X_3. For more than three factors the mathematical approach is necessary. For example, if evaluation of the effect of X_2 is required, this may be done by subtracting the other effects from each of the observations as illustrated in Table 6.14. Similarly, the effects of the other factors can be evaluated. Other methods involve Yates Algorithm [40] and Anova [41]. For this particular example a longer but simpler approach was adopted [14]. Using Table 6.15 as a guide, the average responses are reported on the right. Then the + and – are summed separately and reported on line sum + and sum – respectively. Lastly, the difference between sum + and sum – is divided by the number of + and this value gives the factor effect for any main effect or interaction when two factors are involved.

Before developing a model from the experimental data, the significance of the factors effect and curvature must be determined. If the computed factor effect is larger than the 'minimum factor effect' then one can assume that the

Table 6.14 Factor effect for X_2

X_2	Effect	
$X_2 - X_1$	0.32–0.30 =	0.02
$X_1X_2 - X_1$	1.30–1.21 =	0.09
$X_2X_3 - X_3$	0.62–0.55 =	0.07
$X_1X_2X_3 - X_2X_3$	1.55–1.62 =	– 0.07
	Overall effect	0.11

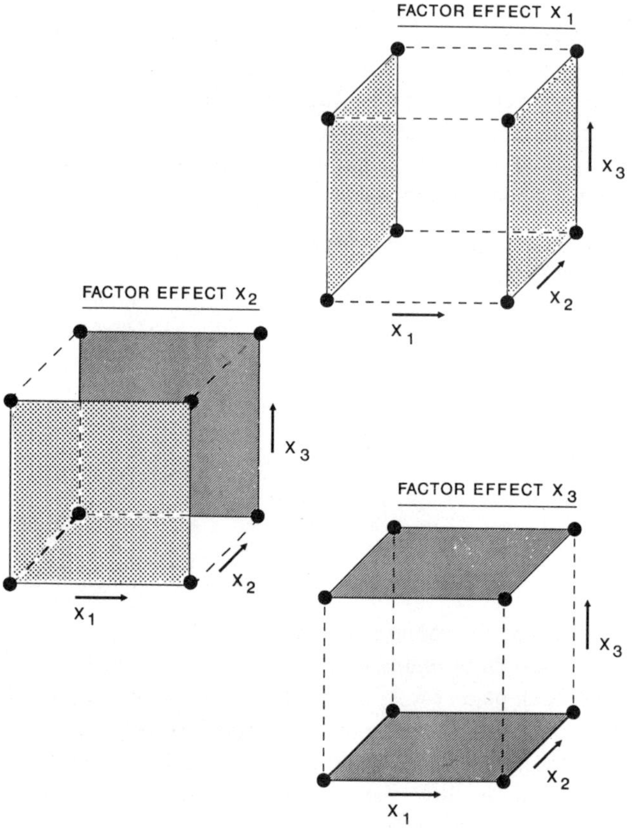

Figure 6.7 Graphical calculation of the main effect.

Table 6.15 Factor effect determination

Trial	Mean	X_1	X_2	X_1X_2	X_3	X_1X_3	X_2X_3	$X_1X_2X_3$	FG
1	+	−	−	+	−	+	+	−	0.30
2	+	+	−	−	−	−	+	+	1.21
3	+	−	+	−	−	+	−	+	0.32
4	+	+	+	+	−	−	−	−	1.30
5	+	−	−	+	+	−	−	+	0.55
6	+	+	−	−	+	+	−	−	1.62
7	+	−	+	−	+	−	+	−	0.62
8	+	+	+	+	+	+	+	+	1.55
Sum +	7.47	5.68	3.79	3.70	4.34	3.79	3.68	3.63	
Sum −	0	1.79	3.68	3.77	3.13	3.68	3.79	3.84	
Overall sum	7.47	7.47	7.47	7.47	7.47	7.47	7.47	7.47	
Difference	7.47	3.89	0.11	−0.07	1.21	0.11	−0.11	0.21	

effect is non zero as is the curvature. The minimum significant factor (MIN) and minimum significant curvature (MIN C) are derived from *t*-test of significance [14].

$$\text{MIN} = ts\sqrt{2/mk}$$
$$\text{MIN C} = ts\sqrt{(1/mk + 1/c)}$$

where:

(MIN)	= minimum significant factor effect,
(MIN C)	= minimum significant curvature,
t	= the value of t at the desired probability level for the number of degrees of freedom in the estimate s,
m	= the number of + signs in the column,
k	= the number of replicates of each trial,
c	= the number of central points,
s	= pooled standard deviation of a single response observation

Having established the significance of the experimental design, an empirical model can be written in the form:

Predicted FG = Grand average + 1/2 the pigment effect of FG

Therefore the model would be:

$$\text{Predicted FG} = 0.81 + 0.49 \times 1 + 0.15 \times 3 + 0.25 \times 1 \times 2 \times 3$$

Many programs are now available to carry out the steps necessary for two-level design experiments.

Although the above is a simple example to illustrate the usefulness of the technique, it can, nevertheless, be seen how each of the factors affect the flocculation gradient and the interaction between them. Many examples have been cited in the literature, where this technique was used with success [42–46]. It has been employed to optimise formulation, to evaluate performance characteristics and to control process conditions.

6.5.5 *Fractional factorial design*

In the previous section it was shown how useful the two-level factorial design is; however, if the factors go beyond six or seven then the number of samples required is too large and becomes unpractical. Plackett and Burman used a fraction of two-level factorial design to estimate the effects of the variables under study with a high degree of confidence. This type of experimental design is very useful when evaluation of many variables is required (4 to 100); it is also known as the screening experimental design. The number of samples required is greatly reduced; for example, to evaluate 19 factors 20 samples are required. However, compared to the full factorial design the fractional factorial design does not give the factor's interaction. Although Plackett and

Burman's [47] experimental designs are able to evaluate up to 99 factors, the most commonly used are designs that study up to 19 factors (see Table 6.16)[14]. Variables can be continuous and discrete.

Analysis of the experimental data is done in same manner as in the previous case with the difference being that for the Plackett–Burman design a ranking of the factor effects is obtained from the largest to the smallest, together with an estimate of the relative importance. Clearly if this type of experimentation is used to screen a series of factors for an estimate of the effect, further optimisation on the variables that exhibit an effect is necessary.

Table 6.16 Twelve–run Plackett–Burman design

Trial	Mean	X_1	X_2	X_3	X_4	X_5	X_6	X_7	X_8	X_9	X_{10}	X_{11}
1	+	+	+	−	+	+	+	−	−	−	+	−
2	+	+	−	+	+	+	−	−	−	+	−	+
3	+	−	+	+	+	−	−	−	+	−	+	+
4	+	+	+	+	−	−	−	+	−	+	+	−
5	+	+	+	−	−	−	+	−	+	+	−	+
6	+	+	−	−	−	+	−	+	+	−	+	+
7	+	−	−	−	+	−	+	+	−	+	+	+
8	+	−	−	+	−	+	+	−	+	+	+	−
9	+	−	+	−	+	+	−	+	+	+	−	−
10	+	+	−	+	+	−	+	+	+	−	−	−
11	+	−	+	+	−	+	+	+	−	−	−	+
12	+	−	−	−	−	−	−	−	−	−	−	−

6.5.6 *Box–Behnken*

There are experiments where the curvature effect has to be estimated in order to establish the degree of interaction. This can easily be achieved by setting a

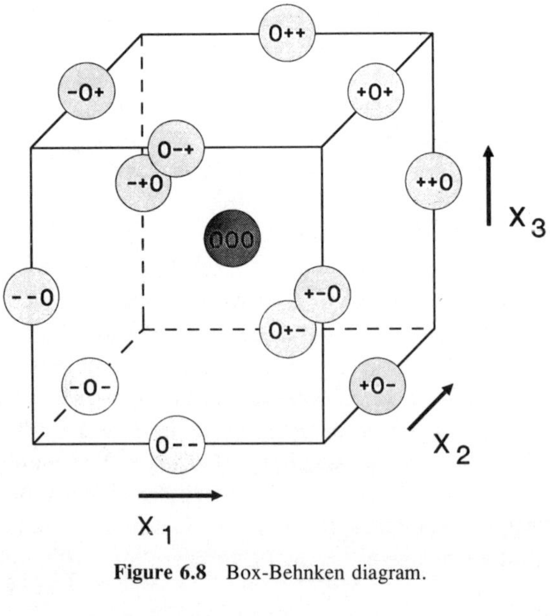

Figure 6.8 Box-Behnken diagram.

three-level factorial experiment (3^3). However, as before, the number of trials can become large when the number of factors (m) increases. Box and Behnken [48] introduced a design which lessened the number considerably. The geometry for a three variable Box–Behnken design is illustrated in Figure 6.8.

All conditions are at the midpoint of the edges, which are equidistant from the centre. For a three-factor, three-level design 27 samples are required, whereas with the Box–Behnken design 15 are sufficient, as illustrated in Table 6.17.

Table 6.17 Three-variable Box–Behnken design

X_1	X_2	X_3	
+1	+1	0	
+1	−1	0	
−1	+1	0	
−1	−1	0	
+1	0	+1	
+1	0	−1	
−1	0	+1	
−1	0	−1	
0	+1	+1	
0	+1	−1	
0	−1	+1	
0	−1	−1	
0	0	0	Centre point
0	0	0	
0	0	0	

Using the Box–Behnken design, Vance [49] developed numerical colour-tolerance regions to determine if a colour sample is acceptable compared to a standard.

6.6 Experimental design with mixtures

In many experimental situations where the factors cannot be varied independently of each other, the proportion of each component must lie between 0 and 1.0 and the sum of the components must be equal to 1. When this condition exists, factorial designs and all the special subsets cannot be used to study the response of multi-component systems.

In recent years researchers have become more aware of the complexity of evaluating multi-component systems and have developed models to study such systems. The early development work was carried out by Claringbold [50] and Scheffe [51, 52]. Other authors such as Thompson and Myers [53], Draper and Lawrence [54] and Snee [55] have developed efficient experimental designs and models for mixtures which clarified the understanding of the applications of mixture designs, as well as the analysis and interpretation of the resulting data. These methods recognise that for systems involving the components by weight or volume they must be either 1 or 100 and be represented by a finite region. For a three- and four-component system the experimental region will be an equilateral triangle and a tetrahedron respectively. The components may be pure or, in the case of coatings, mixtures, in which case they are pseudo-components. Contours of constant response generated by a model fitted to the data can be plotted as a visual aid for interpretation.

6.6.1 *Three-component mixture design*

Many models have been proposed for the analysis of mixture data. The most frequently used is the following special cubic equation derived from the Taylor series [56]:

$$Y = b_1X_1 + b_2X_2 + b_3X_3 + b_{12}X_1X_2 \qquad (1)$$

$$+ b_{13}X_1X_3 + b_{23}X_2X_3 + b_{123}X_1X_2X_3$$

where coefficients are:

$$\begin{aligned}
b_1 &= Y_1 \\
b_2 &= Y_2 \\
b_3 &= Y_3 \\
b_{12} &= 4Y_{12} - 2(Y_1 + Y_2) \\
b_{13} &= 4Y_{13} - 2(Y_1 + Y_3) \\
b_{23} &= 4Y_{23} - 2(Y_2 + Y_3) \\
b_{123} &= 27Y_{123} - 12(Y_{12} + Y_{13} + Y_{23}) + 3(Y_1 + Y_2 + Y_3)
\end{aligned}$$

For the mixture design to be valid, a good approximation of the coefficients, an estimate of the experimental error, and a measure of the lack of fit of the model must be provided.

The coefficients are functions of the measured responses. The experimental error can be determined by replicating the responses. An estimate of the suitability of the model equation is obtained by comparing observed responses versus predicted responses at the threc checkpoint compositions using statistical variance ratios.

Before proceeding, a review of the trilinear coordinate paper and an appraisal of how to plot the composition is helpful. A typical plot is shown in Figure 6.9. The X_1 coordinate, which extends from $X_1 = 0$ at the bottom to $X_1 = 1$ at the vertex, gives the proportion of component 1 in the mixture. The $X_2 = 0$ on the right-hand side to $X_2 = 1$ at the lower left-hand side gives the proportion of component 2 in the mixture. The X_3 coordinate defines the proportion of component 3 in the mixture and extends from $X_3 = 0$ on the left-hand side to $X_3 = 1$ at the lower right. Lines of constant X_1, X_2 and X_3 run parallel to the bottom, right and left side of the triangle, respectively. For

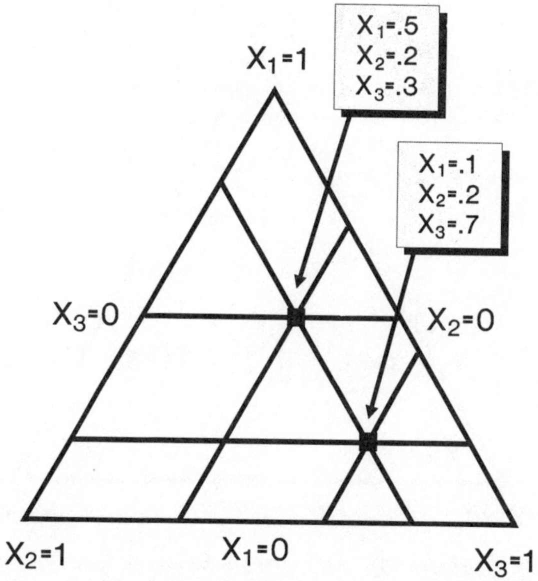

Figure 6.9 Tri-linear co-ordinate plot.

example, the point or composition (0.5, 0.2, 0.3) is the intersection of the line $X_1 = 0.5$, which is halfway between the top and the bottom, the line $X_2 = 0.2$, which is 0.2 of the distance from the right side to the left corner and the line $X_3 = 0.3$, which is 0.3 of the distance from the left side to the right corner. In three-component mixtures $X_1 + X_2 + X_3 = 1$, hence the third coordinate is one minus the sum of the other two. The resulting triangle has only two independent dimensions, and the intersection of any two lines defines a point. For example, the point (0.1, 0.2, 0.7) is the intersection of the lines $X_1 = 0.1$ and $X_2 = 0.2$ or $X_1 = 0.1$ and $X_3 = 0.7$.

For a three-component system, ten examples are required, as given in Table 6.18 and shown graphically in Figure 6.10.

Table 6.18 Three-component simplex design

Components blend	X_1	X_2	X_3	Response Y
1	1	0	0	Y_1
2	0	1	0	Y_2
3	0	0	1	Y_3
4	1/2	1/2	0	Y_{12}
5	1/2	0	1/2	Y_{13}
6	0	1/2	1/2	Y_{23}
7	1/3	1/3	1/3	Y_{123}
8	2/3	1/6	1/6	Y_{1123}
9	1/6	2/3	1/6	Y_{1223}
10	1/6	1/6	2/3	Y_{1233}

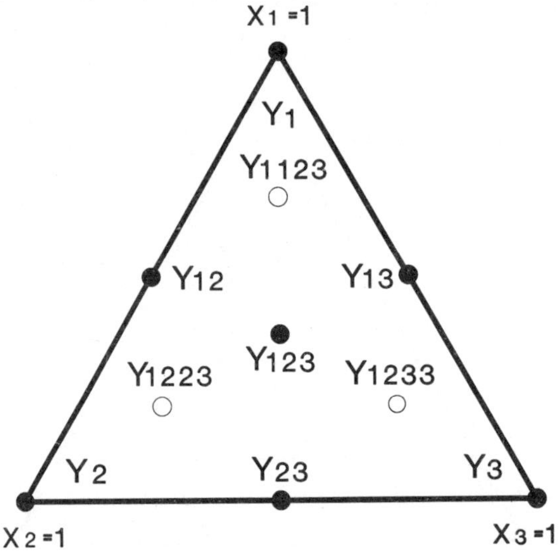

Figure 6.10 Three component mixture design.

The first seven points in the design are used to calculate the coefficients in the special cubic response surface model. The remaining three points (Y_{1223}, Y_{1233} and Y_{1123}) are used to check the accuracy of the prediction of the model. The coefficients in the special cubic model are calculated using equation (1).

Cutrone [57, 58] developed a computer software package COMPLOT III running on an IBM personal computer or any IBM-compatible computer. The software allows the formulator to design mixture experiments, generate contours and, by using the superimposition feature, a ready comparison of contours from different experiments can be made to ensure optimum conditions.

The following four examples explain the design and illustrate the usefulness of the technique.

6.6.1.1 *Development of a semigloss latex paint.* Before setting up the design it is necessary to know which components influence the desired response. In this example it is well known that the levels of binder, pigment and extender (calcium carbonate) are the main parameters that influence the degree of gloss. By first making three paints and then blending these in the appropriate proportions, ten samples were prepared by varying the components by the limits given in the formulation shown in Table 6.19. Table 6.20 enumerates the composition design and the responses.

These results were used to calculate the coefficients in the response surface equation (equation (1)). The resulting coefficients are:

$b_1 = Y_1 = 78.9$

$b_2 = Y_2 = 49.1$

$$b_3 = Y_3 = 15.7$$
$$b_{12} = 4Y_{12} - 2(Y_1 + Y_2) = 4(70.8) - 2(78.9 + 49.1) = 27.2$$
$$b_{13} = 4Y_{13} - 2(Y_1 + Y_3) = 4(51.4) - 2(79.8 + 15.7) = 16.4$$
$$b_{23} = 4Y_{23} - 2(Y_2 + Y_3) = 4(32.7) - 2(49.1 + 15.7) = 1.2$$
$$b_{123} = 27Y_{123} - 12(Y_{12} + Y_{13} + Y_{23}) + 3(Y_1 + Y_2 \times Y_3)$$
$$= 27(53.3) - 12(70.8 + 51.4 + 32.7) + 3(78.9 + 49.1 + 15.7) = 11.4$$

Now that the coefficients in equation (1) have been calculated, the gloss value of any mixture of components X_1, X_2 and X_3 can be determined using the resulting equation:

$$Y = 78.9X_1 + 49.1X_2 + 15.7X_3 + 27.2X_1X_2 + 16.4X_1X_3 + 1.2X_2X_3 + 11.4X_1X_2X_3$$

Table 6.19 Semigloss latex formulation

Material	kg
Propylene glycol	162
Dispersing agent	12
Defoamer	2
Titanium dioxide	290, 200–300
Extender (calcium carbonate)	0, 0–100
Acrylic emulsion	625, 400–650
Water	12
Coalescent	30
Biocide	1
Thickener (2% solution)	110
Defoamer	4
Wetting agent	2

Table 6.20 Compositional design

X_1 (resin)	X_2 (TiO_2)	X_3 (extender)	Y(60°gloss)
650.0	200.0	0.0	78.9
400.0	300.0	0.0	49.1
400.0	200.0	100.0	15.7
525.0	250.0	0.0	70.8.
525.0	200.0	50.0	51.4
400.0	250.0	50.0	32.7
483.0	233.3	33.3	53.3
566.7	216.7	16.7	66.4
441.7	266.7	16.7	48.9
441.7	216.7	66.7	36.7

To carry out these time-consuming calculations and plot the desired contours, COMPLOT III was used, which first calculates all the coefficients, and then, by using equation (1), generates all the isoresponse curves that are required. The response surface contour plot for gloss is shown in Figure 6.11. If blends are made along any of the isoresponse curves, the same value of gloss is obtained. For example, if a gloss value is desired, there is an infinite number of possibilities for achieving this. If point A was chosen, the

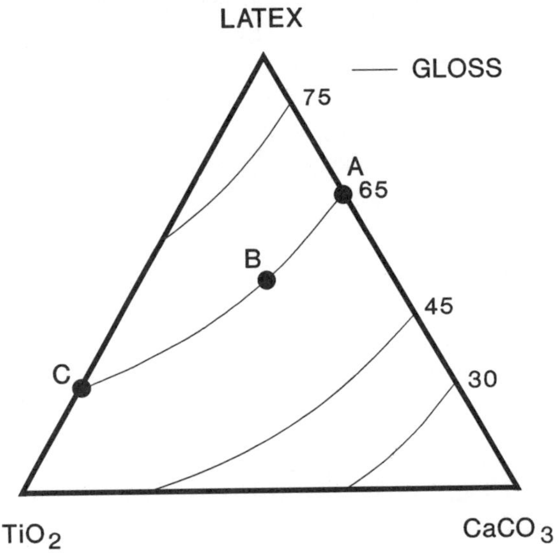

Figure 6.11 Contour plot for 60° gloss.

proportion would be X_1 (latex) = 0.65, X_2 (TiO_2) = 0 and X_3 (extender) = 0.35. If point B was chosen, the proportion would be X_1 (latex) =0.50, X_2 (TiO_2) = 0.25 and X_3 (extender) = 0.25. Lastly, if point C was chosen, then the proportions would be X_1 (latex) = 0.22, X_2 (TiO_2) = 0.78 and X_3 (extender) = 0. If component X_1 is the most expensive, the formulator would choose point C where the least amount of component X_1 is used.

6.6.1.2 *Development of a high opacity semigloss latex paint.* In the first example only one response was measured. In this example the paint has to have high opacity (contrast ratio) and the same value for gloss as in the previous example. Using the same three components and the same experimental design, gloss and contrast ratio for 25 μm thick films were measured. The response surface contour plot for contrast ratio is shown in Figure 6.12. The contrast ratio responses are the broken lines. It can be seen from Figure 6.12 that a gloss of 65 units at a relatively high contrast ratio (94.0%) is obtained at point C where the proportions of the components are X_1 (latex) = 0.22, X_2 (TiO_2) = 0.78 and X_3 (extender) = 0.

6.6.1.3 Development of a matt latex paint. In this example the aim was to develop a matt latex paint with high contrast ratio, good scrub resistance and relatively low porosity, at a fixed cost per litre of pigmentation. The formulation used is given in Table 6.21.

The pigmentation level was kept constant at 55% pigment volume concentration. The three components were clay, titanium dioxide and calcium carbonate. The responses measured were: contrast ratio at 25 μm, cost, ASTM

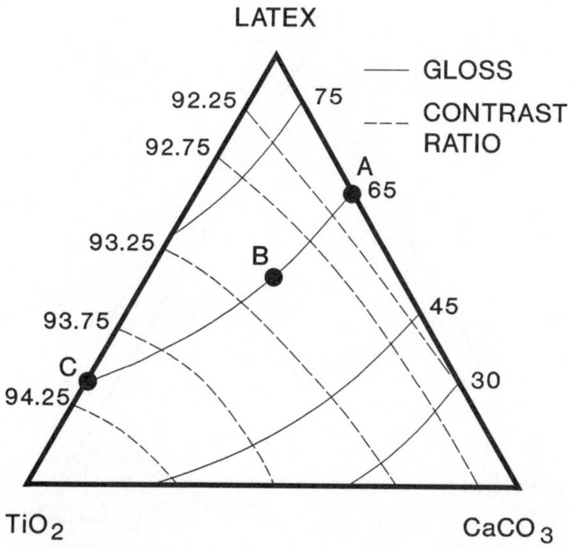

Figure 6.12 Contour plot for 60° gloss and opacity.

Table 6.21 Matt latex formation

Material		kg
Water		140.2
Dispersing agent		2.2
Wetting agent		4.4
Defoamer		2.0
Ethylene glycol		30.9
Pigmentation :	titanium dioxide	
	clay	179 litre
	calcium carbonate	
Thickener (2% solution)		78.0
Water		23.2
Defoamer		1.8
Thickener (2% solution)		219.4
Coalescent		6.9
Biocide		1.0
Water		30.0
Latex		308.7

porosity and scrub resistance. The response surface contour plots are shown in Figures 6.13, 6.14, 6.15 and 6.16.

It can be seen by comparing Figures 6.13 and 6.14 that for a contrast ratio of 96% the cost can vary from approximately 0.60 to more than 1.00 £/litre of pigmentation. If 0.70 £/litre is selected as the cost, there are two possible formulations, as shown in Figure 6.17. The composition at point X is X_1 (clay) = 0.70, X_2 (TiO_2) = 0.30 and X_3 (calcium carbonate) = 0, whereas the composition at point Y is X_1 = 0.27, X_2 = 0.33 and X_3 = 0.40. If opacity and cost were the only requirements, formulating at either point would give the same

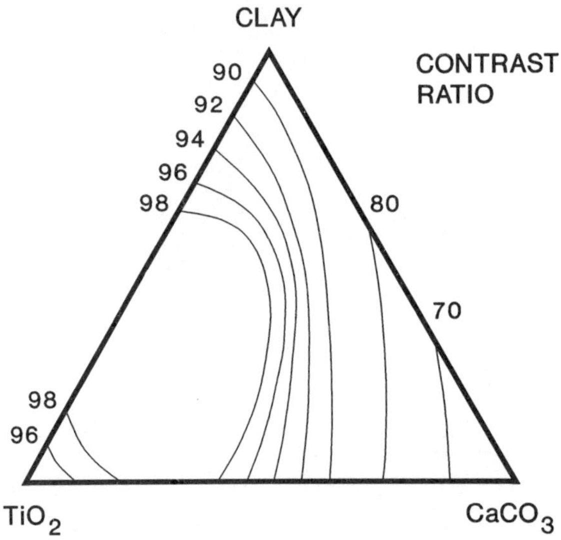

Figure 6.13 Contour plot for opacity.

results. However, while for a matt latex paint a high contrast ratio is very important, equally important is the film porosity. At point X the film is relatively porous and has low scrub resistance (see Figure 6.18), whereas at formulating point Y (Figure 6.19) the film is much less porous and has better scrub resistance. Hence, by making ten samples, a matt latex formulation has been developed with optimum opacity and film integrity at a predetermined cost.

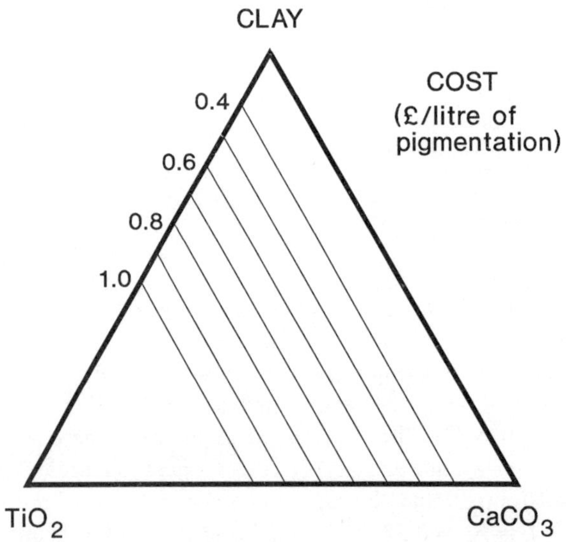

Figure 6.14 Contour plot for cost.

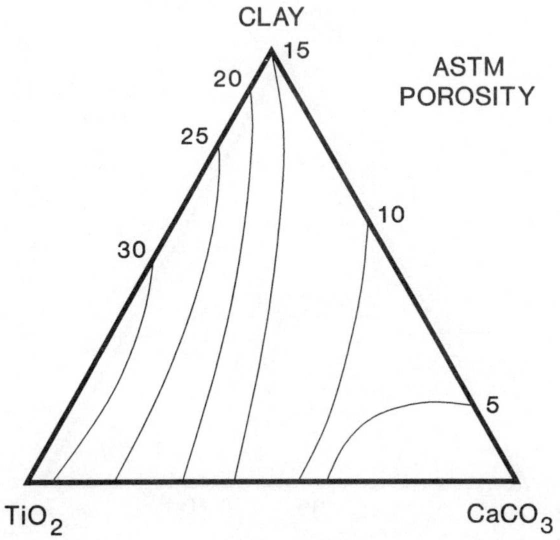

Figure 6.15 Contour for ASTM porosity.

6.6.1.4 *Evaluation of alternative raw materials.* Another area where the three-components contour plotting technique is of great importance is in the evaluation of alternative raw materials. This is a process that paint formulators perform regularly for a variety of reasons such as limited supply, economics or performance of the specific raw material under evaluation.

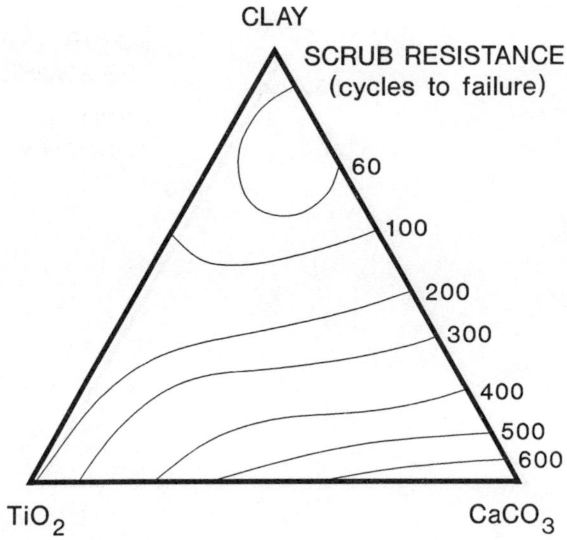

Figure 6.16 Contour for scrub resistance.

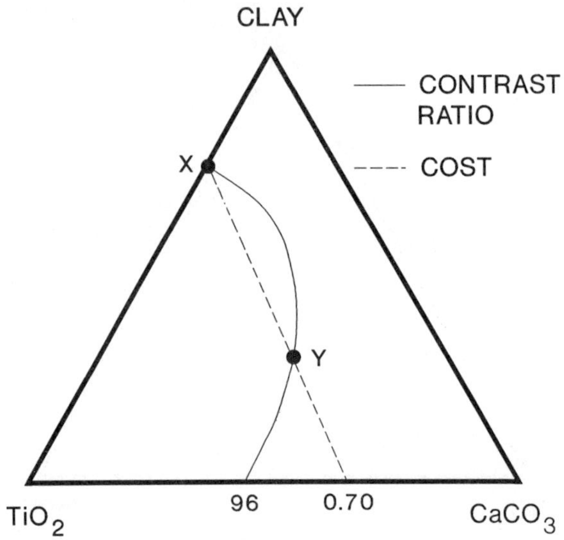

Figure 6.17 Contour plot for opacity at 96% contrast ratio and cost at 0.70 £/litre.

In this example the existing formulation, as described in Table 6.22, had to be modified in order to reduce the raw material cost while maintaining the same performance properties. The properties of particular interest in this example were 60° gloss (40) and the level of opacity (contrast ratio = 94.25%) at a raw material cost less than 0.67 £/litre.

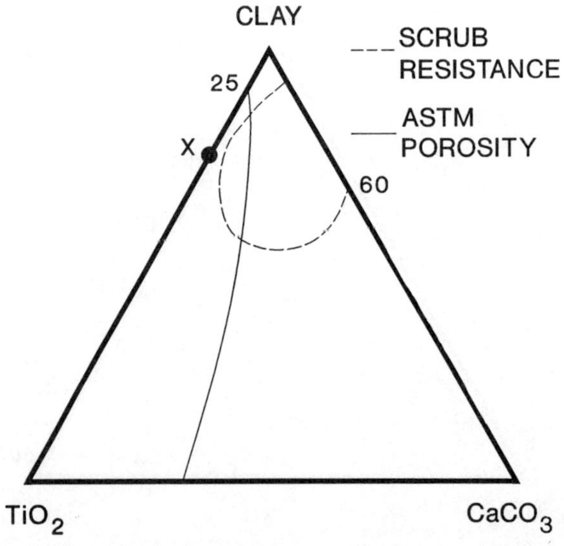

Figure 6.18 Contour plot for a scrub resistance of 60 cycles and an ASTM porosity of 25.

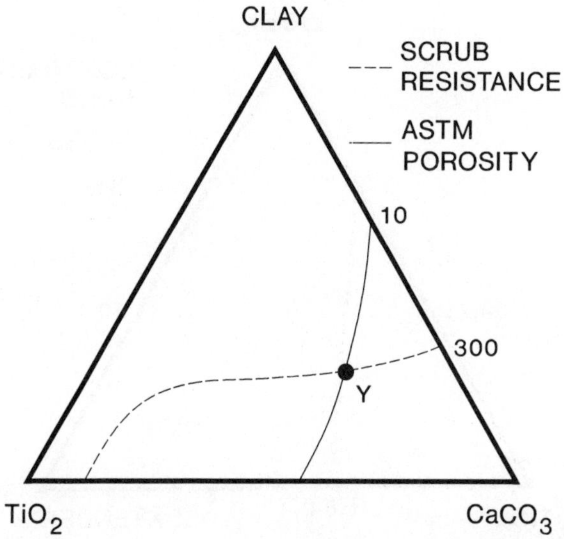

Figure 6.19 Contour plot for a scrub resistance of 300 cycles and an ASTM porosity of 10.

Two extenders were evaluated: A and B. The limits of the design (see Table 6.23) were chosen in such a way that the desired levels of gloss and opacity are in the central region of the triangle. Superimposing the contour responses for extender A, it can be seen from Figure 6.20 that the desired level of gloss and opacity is obtained at a cost of 0.64 £/litre. Repeating the evaluation with

Table 6.22 Semigloss latex formulation

Material	kg
Water	80
Propylene glycol	80
Dispersing agent	10
Defoamer	2
Titanium dioxide	254
Extender	26
Acrylic emulsion	450
Water	23
Coalescent	25
Biocide	1
Thickener (2% solution)	112
Defoamer	4
Wetting agent	2

Table 6.23 Limits of the experimental design

Components	Minimum (kg)	Maximum (kg)
Latex	400	650
Titanium dioxide	200	300
Extender	0	100

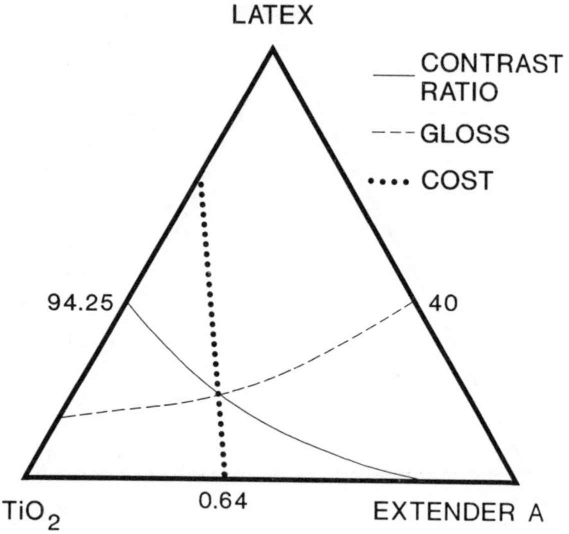

Figure 6.20 Contour plot for opacity, gloss and cost for extender A.

extender B, the same level of performance as the standard can be achieved at a cost of 0.62 £/litre (Figure 6.21).

To ensure that the predicted values were correct, a sample was prepared at the optimum point (latex 11%, TiO_2 = 51% and extender = 38%) and the results are as shown in Table 6.24.

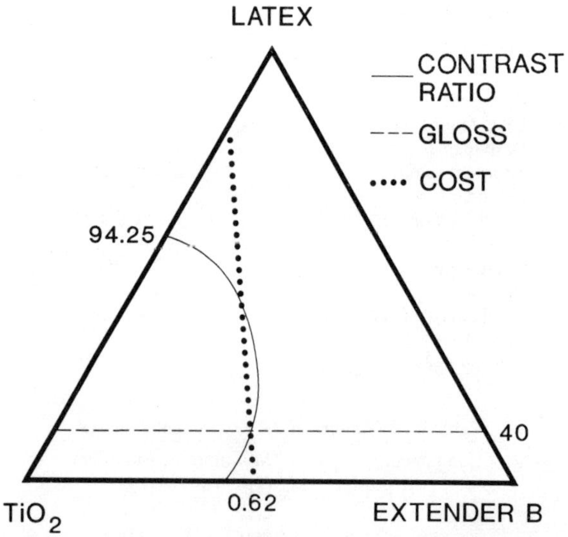

Figure 6.21 Contour plot for opacity, gloss and cost for extender B.

Table 6.24 Predicted versus observed results

	Predicted	Observed
60° gloss	40	38
Contrast ratio	94.25%	94.07%

6.6.2 *Four-component mixture design*

For the four component mixture design the model equation is the same as used previously except it has more terms included for the fourth component:

$$Y = b_1X_1 + b_2X_2 + b_3X_3 + b_4X_4 + b_{12}X_1X_2 + b_{13}X_1X_3 + b_{23}X_2X_3 + b_{24}X_2X_4 + b_{34}X_3X_4 + b_{123}X_1X_2X_3 + b_{124}X_1X_2X_4 + b_{134}X_1X_3X_4 + b_{234}X_2X_3X_4$$

where coefficients are:

$$b_1 = Y_1 \qquad b_{23} = 4Y_{23} - 2(Y_2 + Y_3)$$
$$b_2 = Y_2 \qquad b_{24} = 4Y_{24} - 2(Y_2 + Y_4)$$
$$b_3 = Y_3 \qquad b_{34} = 4Y_{34} - 2(Y_3 + Y_4)$$
$$b_4 = Y_4 \qquad b_{123} = 27Y_{123} - 12(Y_{12} + Y_{13} + Y_{23}) + 3(Y_1 + Y_2 + Y_3)$$
$$b_{12} = 4Y_{12} - 2(Y_1 + Y_2) \qquad b_{124} = 27Y_{124} - 12(Y_{12} + Y_{14} + Y_{24}) + 3(Y_1 + Y_3 + Y_4)$$
$$b_{13} = 4Y_{13} - 2(Y_1 + Y_3) \qquad b_{134} = 27Y_{134} - 12(Y_{13} + Y_{14} + Y_{34}) + 3(Y_2 + Y_3 + Y_4)$$
$$b_{14} = 4Y_{14} - 2(Y_1 + Y_4)$$

As with the three-component, the four key components are selected. The concentration of these components is then varied within predetermined limits, and the influence of this variation in composition upon properties determined. Having selected the components whose concentration is to be varied, the experimental compositions are then determined according to the scheme outlined in Table 6.25. The first fourteen points are used to calculate the coefficients in the special cubic model equation, while the remaining five points can be used to verify the accuracy of the data produced.

It has already been stated that in four-component mixture design experiments the area of experimentation is described by a tetrahedron. As studying surface response contours over a three-dimensional area is difficult, a simplified optimisation routine was developed. A 'slice' can be extracted at any composition within the tetrahedron. This is performed by holding the level of one component constant and monitoring the variation in response that results from changes in the relative concentrations of the other three components. This is illustrated in Figure 6.22 where component X_1 is fixed at 0.5 (i.e. 50%). This resulted in the experimentation area being reduced to an equilateral triangle (in effect becoming a three-component experiment). An alternative method is to hold the concentration of component X_1 at its maximum level and thus represent the experimentation area by four discrete equilateral triangles, as shown in Figure 6.23. The distribution of the response values within the two-dimensional experimentation area is shown in Figure 6.24. By employing a special cubic equation, the software generates continuous contours from these isolated data points. By studying the surface contours, a composition having the required properties can be selected. The composition of any point within the tetrahedron can be determined by simple geometric construction.

Table 6.25 Four component simplex design

Components blend	X_1	X_2	X_3	X_4	Response Y
1	1	0	0	0	Y_1
2	0	1	0	0	Y_2
3	0	0	1	0	Y_3
4	0	0	0	1	Y_4
5	1/2	1/2	0	0	Y_{12}
6	1/2	0	1/2	0	Y_{13}
7	1/2	0	0	1/2	Y_{14}
8	0	1/2	1/2	0	Y_{23}
9	0	1/2	0	1/2	Y_{24}
10	0	0	1/2	1/2	Y_{34}
11	1/3	1/3	1/3	0	Y_{123}
12	1/3	1/3	0	1/3	Y_{124}
13	1/3	0	1/3	1/3	Y_{134}
14	0	1/3	1/3	1/3	Y_{234}
15	1/4	1/4	1/4	1/4	Y_{1234}
16	5/8	1/8	1/8	1/8	Y_{11234}
17	1/8	5/8	1/8	1/8	Y_{12234}
18	1/8	1/8	5/8	1/8	Y_{12334}
19	1/8	1/8	1/8	5/8	Y_{12344}

The usefulness of four-component mixture design experiments is illustrated by using the technique to optimise actual latex paint formulations [59].

6.6.2.1 *Development of a semigloss latex paint.* There is currently a great deal of interest in modifying the rheological behaviour of latex paints. This is largely due to the successful commercial development of associative thickeners [60]. Although offering significant advantages over traditional cellulosic thickeners, such as the elimination of roller spatter, increased high-shear viscosity and improved levelling, some difficulties have been reported in the use of associative thickeners [61, 62]. This has often led the practical paint formulator to employ mixtures of thickeners. Thus, a mixture design experiment was devised to determine the optimum concentrations of three rheological additives and an anionic surfactant in a semigloss latex paint.

A water-soluble polyurethane associative thickener was used as the principal rheology modifier in the system. Such thickeners contain pendant hydrophobic groups capable of undergoing intermolecular 'associations', and also of absorbing onto the surfaces of dispersed particles (primarily latex and pigment). While their performance is affected by parameters such as latex particle size and the presence of co-solvents, the introduction of surfactants can also have

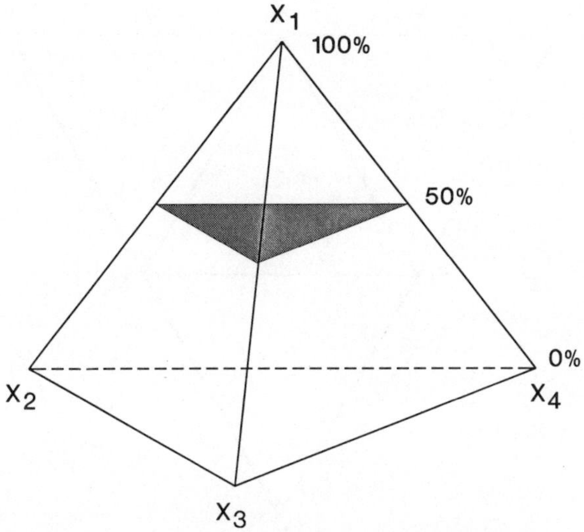

Figure 6.22 Four component experimental region.

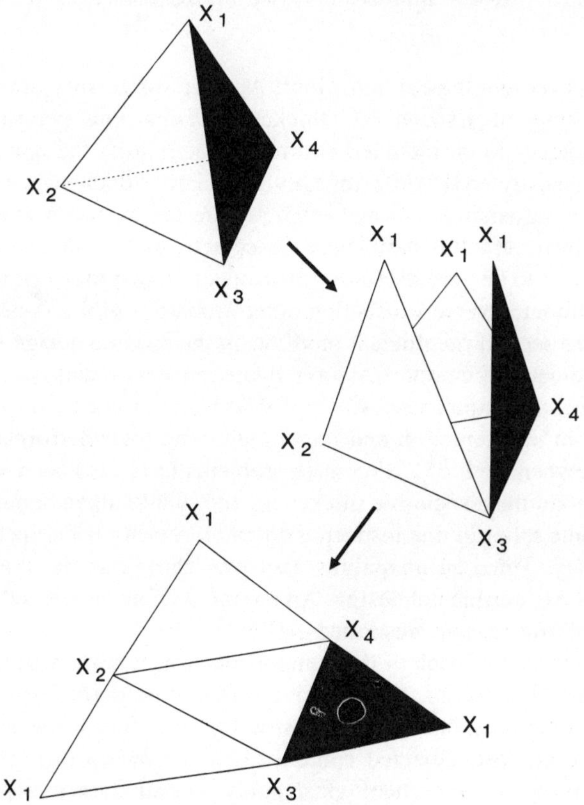

Figure 6.23 Tetrahedron manipulation.

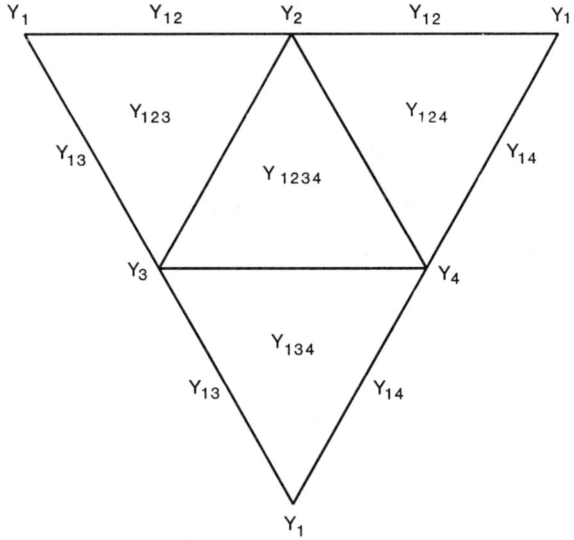

Figure 6.24 Response distribution within the four component experimental area.

a major effect on rheological behaviour. Anionic surfactants can depress the Stormer viscosity of associatively thickened paints, thus permitting higher levels of thickener to be included and hence facilitating the development of high-shear viscosity [63]. Unfortunately associative thickeners tend to give rather poor sag resistance [64] and often require the inclusion of an auxiliary thickener to overcome this deficiency. Since activated alkanolamine titanates have been found to be very useful in this respect, in that they can improve sag resistance without adversely affecting other properties of the system, one was selected as the second rheological modifier in the mixture design experiment. These organometallic complexes form hydrogen bonds with suitable latices and in this manner impart viscosity and thixotropy to latex paints. Details of the mechanism of interaction and factors affecting their performance can be obtained elsewhere [64, 65]. Syneresis problems have also been encountered in paints containing associative thickeners and, while alkanolamine titanates may offer some relief in this respect, a common remedy is the inclusion of an attapulgus clay. Thus, an attapulgus clay was chosen as the third rheology modifier in the experimental design. An anionic surfactant was selected as the fourth variable for reasons described earlier.

The concentration of each of the components selected was varied within the limits shown in Table 6.26. An extensive series of properties was determined for the resultant paints, including viscosity, sag resistance, flow and levelling, and raw material cost. Selected contour plots are shown in Figures 6.25 to 6.28. In carrying out the rheological study, certain target responses were selected, as shown in Table 6.27. In order to determine compositions having

Table 6.26 Experimental design parameters

Components	Description	Composition(kg/1000 l)	
		Minimum	Maximum
X_1	Attapulgus clay	16.0	0.0
X_2	Associative thickener (urethane)	36.0	16.0
X_3	Anionic surfactant	12.0	0.0
X_4	Activated alkanolamine titanate	8.0	0.0

Table 6.27 Target parameters

Property	Range
High shear viscosity	1.40–1.50
Flow and levelling	7.5–8.5
Sag resistance	9.5–10.5
Stomer viscosity	85–95

these target properties the following optimisation procedure was employed. Initially the contours representing compositions having the required properties were highlighted and superimposed on a single plot (Figure 6.29). In this manner a range of compositions having the properties required was indicated. For clarity, only contours having mid-point levels of the properties under examination are given in Figure 6.29. In actual experiments, contours describing the full range of desired properties are generated on transparencies and then superimposed. Since the two-dimensional plot was achieved by setting the concentration of component X_1 at its minimum, additional plots were produced by increasing the concentration of this component (Figures 6.30a–e). Increasing the concentration of component X_1 resulted initially in an increase in the range of composition having the required properties. At $X_1 = 0.10$, the area of composition having the target properties extends beyond a maximum, until $X_1 = 0.20$ where the target properties can no longer be achieved. Thus, having determined the limits of compositions having the required properties, the optimum composition was selected as that which met the target properties at the lowest raw material cost. Details of this composition are given in Table 6.28.

To determine the validity of the experimental design, the optimum composition was prepared and its actual properties compared with the predicted

Table 6.28 Optimum composition

Component	Optimum level of component (%)	Weight (kg)
X_1	10.0	1.60
X_2	16.8	19.36
X_3	9.7	1.16
X_4	63.5	5.08

Table 6.29 Replicated and predicted responses (semigloss latex)

Property	Predicted	Observed
High-shear viscosity	1.47	1.39
Flow and levelling	8.0	7.5
Sag resistance	10.0	9.5
Stomer viscosity	88.0	87.0

values. The results obtained, as shown in Table 6.29, indicate that the predicted and actual responses correlate well, the difference between the values being within the limits of the test methods employed.

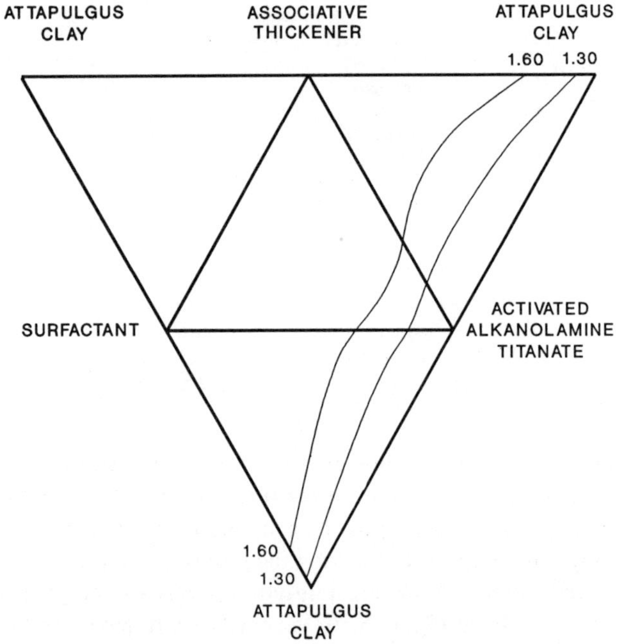

Figure 6.25 Contour plot for high-shear viscosity.

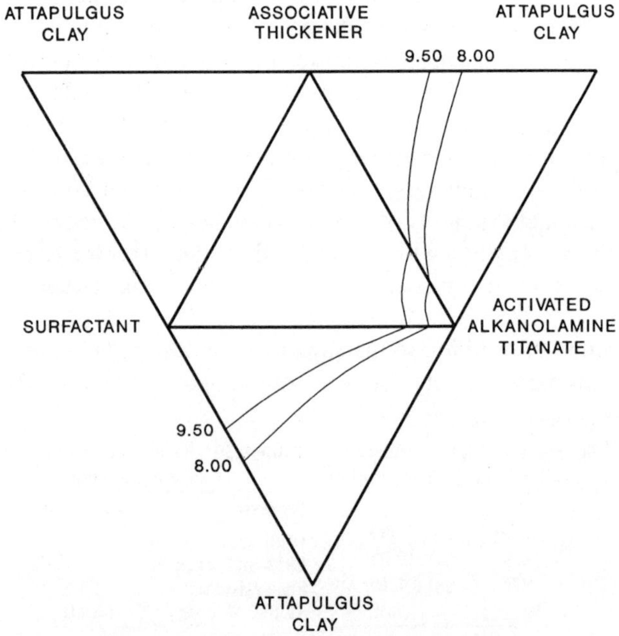

Figure 6.26 Contour plot for flow and levelling.

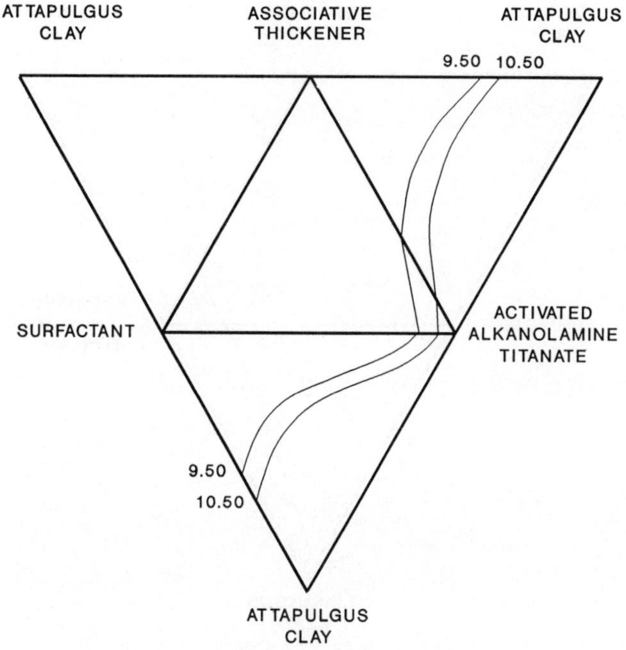

Figure 6.27 Contour plot for sag resistance.

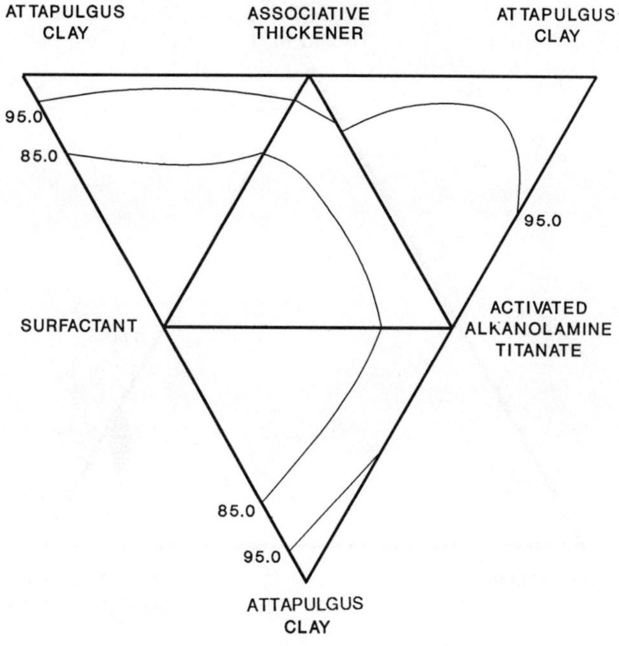

Figure 6.28 Contour plot for Stormer viscosity.

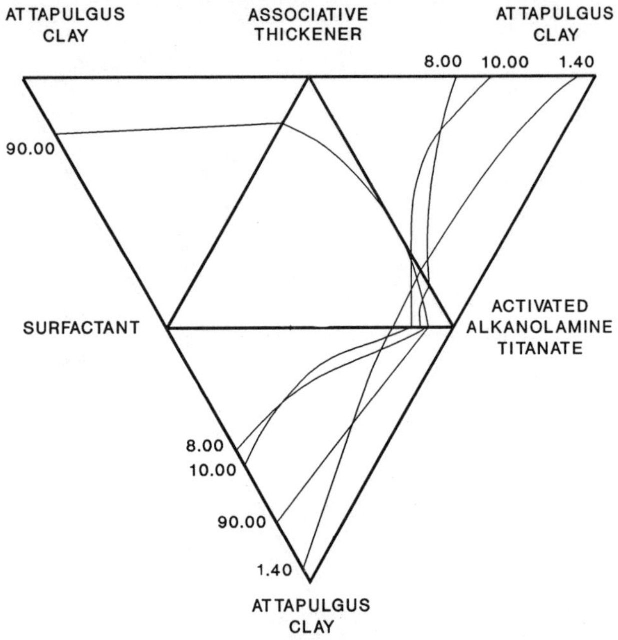

Figure 6.29 Superimposition plot of target properties.

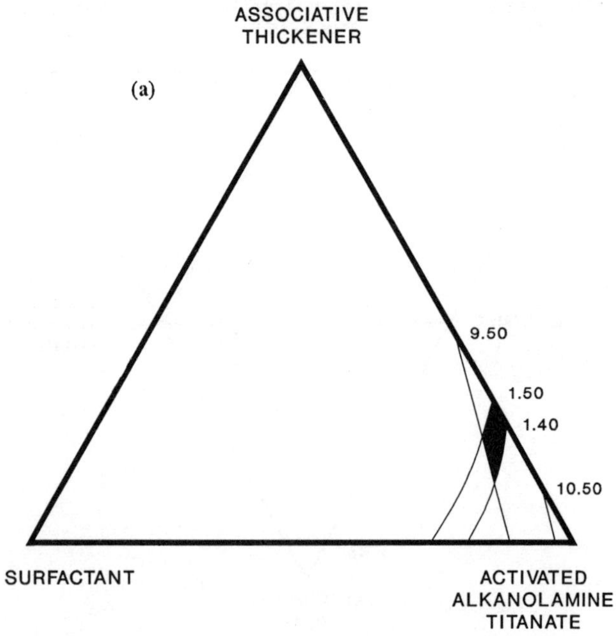

Figure 6.30 (a) 0% pseudocomponent X_1; (minimum concentration).

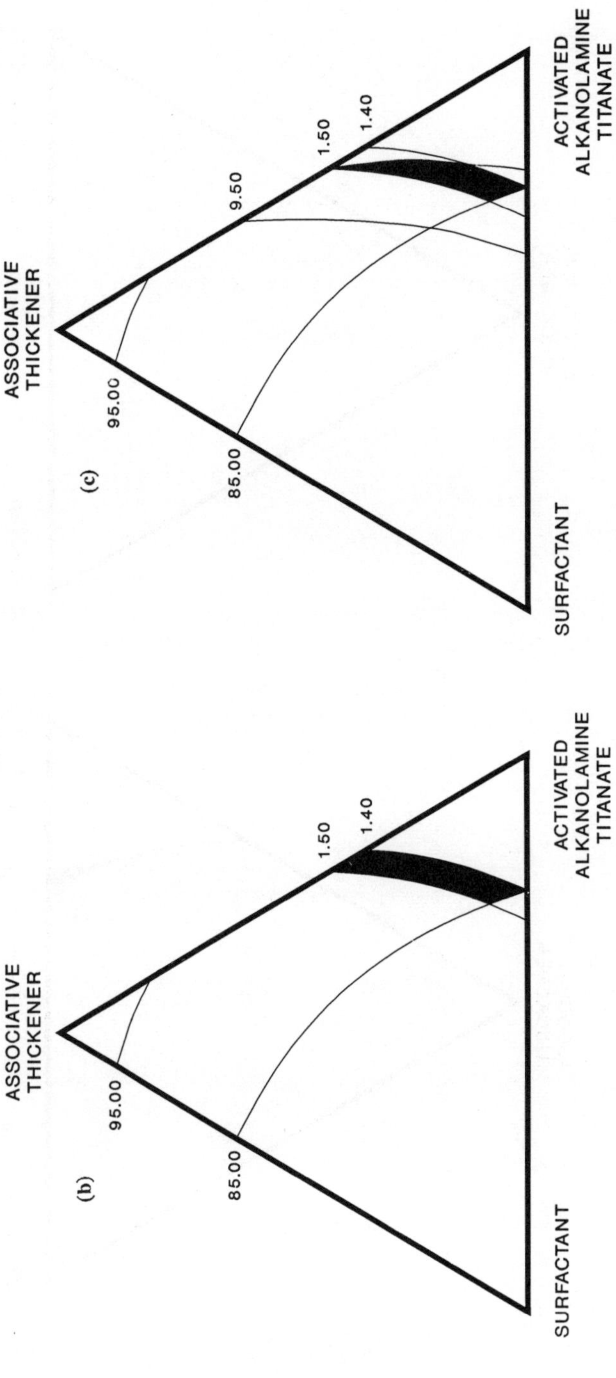

Figure 6.30 (b, c) (b) 10% pseudocomponent X_1; (c) 15% pseudocomponent X_1.

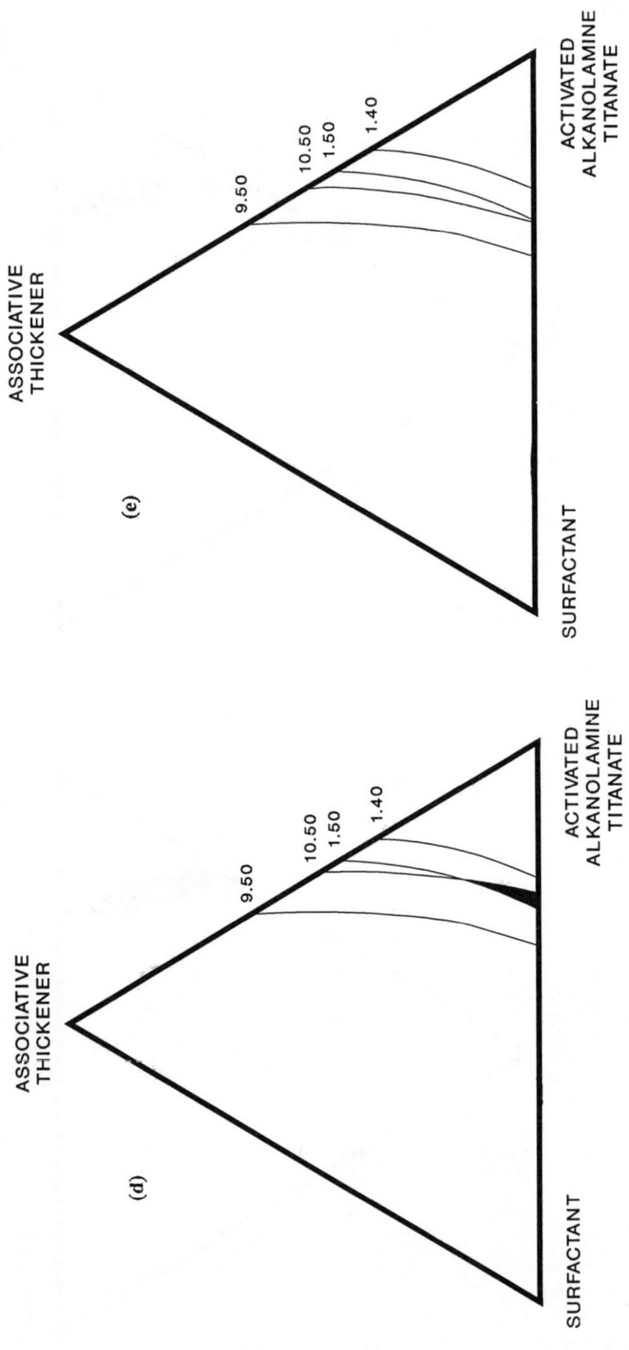

Figure 6.30 (d, e) (d) 18% pseudocomponent X_1; (c) 20% pseudocomponent X_1.

6.6.3 *Simplex mixture screening design*

In Section 6.1 the use of simplex latex lattice design experimentation was discussed in relation to mixtures with few components under study. In the normal course of experimental investigation, however, the researcher would begin with a relatively large number of variables. Hence, the first step would be to study all potentially important variables in a screening study.

The technique [14, 52, 66–68] for the design of simplex lattice mixture studies can apply to simplex screening studies. The simplex mixture screening design is a subset of the full simplex lattice design. The composition of the blends in the simplex screening design is achieved by the use of a $(3q + 1)$ design, which contains the q vertices, or pure components; the q interior blends; the overall centroid; and the q end effect blends, where q is the number of components. Analysis of the simplex mixture screening design is divided into two parts: graphical and mathematical. The graphical part consists of plotting the responses along the component axes and drawing lines between each response associated with each component. The mathematical analysis of mixture screening design involves four sets of calculations: the component mixture model coefficients (bi's), the component effect (Ei's), the confidence limits for the coefficients and the confidence limits for the effects.

This technique was used by Hesler *et al.* [69] to evaluate the effects of 12 dispersing and stabilising surfactants in an emulsion paint.

6.7 Theoretical approach to formulation

The previous sections have dealt with useful and elegant methods to optimise existing formulations and develop new ones; however, the formulator's ultimate aim is to have a universal formulation concept, i.e. a way to predict which raw material should be used and the type of dispersing machinery necessary to obtain optimum results for a coating system. To achieve this, fundamental knowledge of interactions between pigment, extender, polymer, solvent and other paint additives expressed by means of the acid-base concept, flocculation theory and rheology is necessary.

In the past two decades, several authors have approached the task of theoretical formulation and the prediction of paint performance. Bierwagen [70] developed a computer-based method of formulation and cost minimisation based on the prediction of a latex coating critical pigment volume concentration (CPVC) and the use of the pigment volume concentration (PVC)/CPVC ratio. The mathematical model for analysing CPVCs in a mixture of multicomponent pigments is based on pigment and emulsion particle size and oil absorption (absorbed layer). This model was refined by taking into account the packing characteristics of the pigment, extenders and emulsion. Using the three-component contour technique, Cutrone *et al.* [71] studied the interactions between pigments and different types of rheology modifiers. Stieg [72] also developed an empirical model describing the pigment and extender distribution in a paint film.

Recently Simpson *et al.* [73] developed a mathematical model for predicting the state of pigment dispersion in paint films. The model uses the Monte Carlo method to generate random positions in a box for spheres with specified size

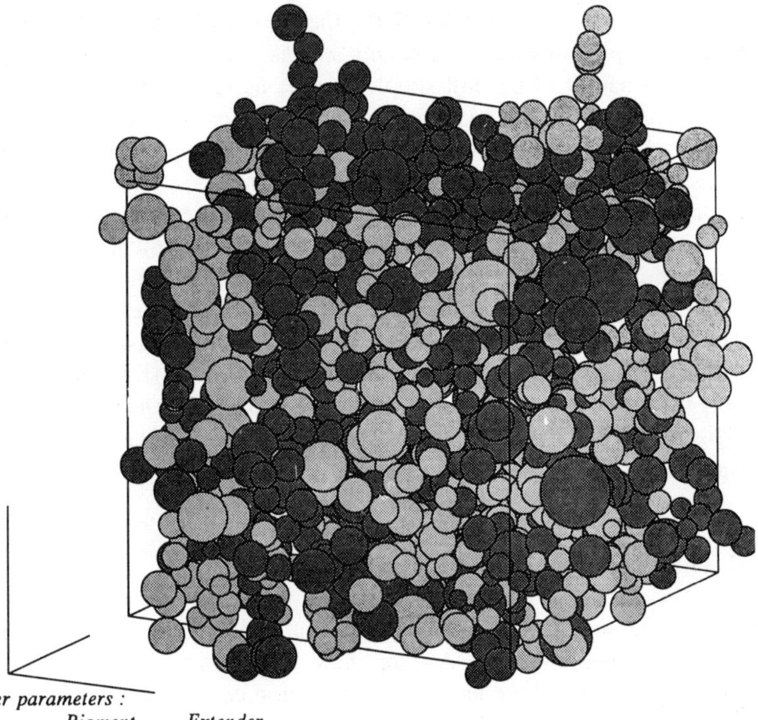

Number parameters :		
	Pigment	*Extender*
Mean	0.1790	0.1790
Actual	0.1738	0.1753
S.Dev	1.3000	1.3000
Actual	1.2894	1.2879
Number	612	591
Weight Parameters :		
	Pigment	*Extender*
Mean	0.2200	0.2200
Actual	0.2144	0.2177
S.Dev	1.3000	1.3000
Actual	1.3182	1.3318
Concentrations :		
	Pigment	*Extender*
Vol%	15.0000	15.5000
Actual	13.0017	12.9545
NNdist	0.0093	0.0072

Figure 6.31 Dispersion simulation of a paint film containing equal parts of pigment and extender (0.22 μm).

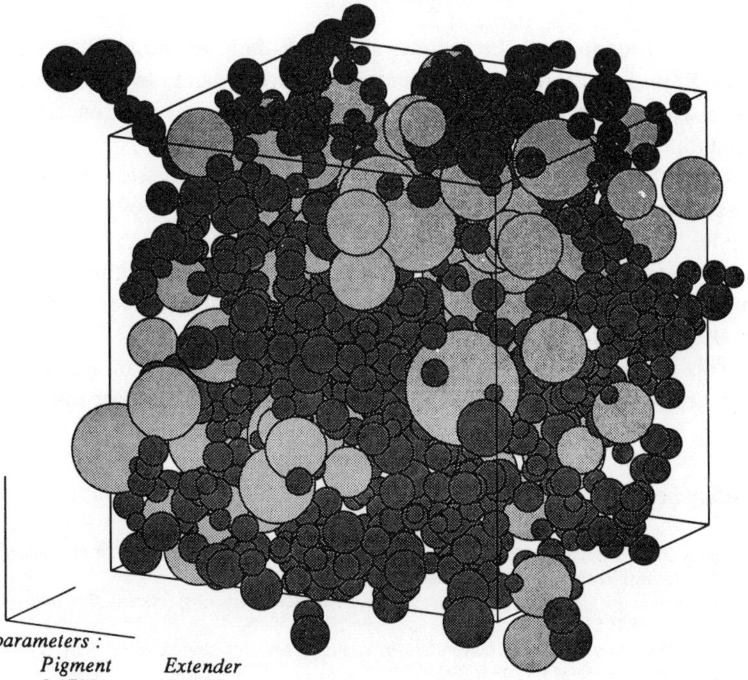

Number parameters :

	Pigment	*Extender*
Mean	0.1790	0.4067
Actual	0.1767	0.4089
S.Dev	1.3000	1.3000
Actual	1.2844	1.2676
Number	869	76

Weight Parameters :

	Pigment	*Extender*
Mean	0.2200	0.5000
Actual	0.2174	0.4864
S.Dev	1.3000	1.3000
Actual	1.3171	1.2768

Concentrations :

	Pigment	*Extender*
Vol%	9.6200	14.9500
Actual	12.5615	13.0385
NNdist	0.0062	0.0277

Figure 6.32 Dispersion simulation of a paint film containing equal parts of pigment and extender (0.50 μm).

distribution and designed volume concentration. An indication of the degree of dispersion is obtained by calculating the nearest neighbour distance for each sphere and the distance distribution. The model which accounted for pigment flocculation according to a Smoluchowski distribution can predict the effects the extender particle may have on pigment dispersion.

The technique used belongs to a class known as the Monte Carlo method. In the study, a geometrical description is built up of a typical test volume of the paint film. The model is generated from the known characteristics of the film constituents and the manner in which they are brought together in the film. Representation of the elements (i.e. particles and cluster of particles) and their disposition relative to one another is achieved by generating pseudo random numbers.

Once the geometrical picture has been formed, statistical measures of the dispersion are calculated and taken to be representative of the paint film as a whole.

An example of the dispersion simulation is shown in Figures 6.31 and 6.32 where two paint films of equal PVC but different extender particle size are evaluated for dispersion.

This is a very powerful technique because the model can account for packing, crowding, flocculation (hetero and homo) and any other interaction.

References

1. New York Society for Coatings Technology, Utilization of computers in the coating industry, *J. Coatings Tech.* **56**. (719) (1984) 77.
2. J.P. Kennedy, Computerized process control in manufacturing process, *J. Coatings Tech.* **55**. (706) (1983) 51.
3. D.T. Wu, Application of computers in coatings research and development. *XVIII FATIPEC Congress*, Venezia **3**. (1986) 393.
4 R. Zum-Bruch, Computer-controlled formulation of lacquers with particular reference to calculation of pollutants, *Farbe Lacke*. **90**. (11) (1984) 942.
5. C.C. Grau, Computer software for classifying hazardous material, *Farbe Lacke*. **89**. (5) (1983) 377.
6. Anon, Computer delineates profitability, tightens inventory control. *American Inkmaker* **59**. (1) (1981) 28.
7. B. Kneller, Computerized formulation management and automated control in paint manufacture, *Paint and Resin* (June 1987). p. 38.
8. Marcus Paint Co, Computer Management system removes production roadblocks, *Modern Paint and Coatings*. (July 1988) p. 41.
9. L. Vincentz, La fonctionnalité. sans fioritures. *European Coatings Journal* **5** (1989) 444.
10. D.R. Duncan, The colour of pigment mixtures, *Proc. Phys. Soc. Lond.* **52** (1940) 380.
11. D.R. Duncan, The colour of pigment mixtures, *J. Oil and Colour Chemists Assoc*. **32**. (1949) 296.
12. R. Davidson, and H. Hemmendinger, Tristimulus difference computer, *J. Optics Soc. Am.* **51** (1961) 1459.
13. G. Gruppi, Measurements of appearance of paints: instruments and their use, *Verniciatura Industriale* **22** (Sept 1989) 257, 277.
14. E.I. Du Pont de Nemours & Co. *Strategy of Experimentation* (October 1975) Wilmington, USA.
15. H. Grassman, Theorie der Farbenmichung, *Poggendorffs Ann.* **89**. (1853) 69.
16. J. Guild, The colorimetric properties of the spectrum, *Phil. Trans. R. Soc.* A **230** (1931) 149.
17. W.D. Wright, The colorimetric properties of the spectrum. *Philo. Trans. R. Soc.* A **230** (1931) 149.
18. K. Schlapfer, Comparaison des appareils de mésure des couleurs, *European Coatings Journal* **5**. (1989) 388.
19. D.M. Williams, and V.L. Bacchetta, Computer calculations of paint formulations, *J. Paint Tech.* **39** (508) (1967) 267.
20. Philadelphia Society for Paint Technology, The computer and formulation design. *J. Paint Tech.* **41** (1969) 528.40.
21. General Electric Bulletin PAINT$ program.

22. F. Biglieri, La matematica nelle pitture, *L'industria della Vernice* **13** (8) (1982) 38.
23. S.E. Kendall, Paint lab calculations on pocket computers, American Paint and Coating Journal, **18** (Feb. 1985) 55.
24. K. Hodgson, Micro computer costing, stock and production aids, *Polymer Paint Colour Journal* **174** (4125) (1984) 584.
25. A.D. Sheen, BP solves your reformulation problem, *Polymers Paint Colour Journal* **178**. (4207) (1988) 70.
26 J.D. Hall, Implementation of laboratory paint formula calculation on an Apple II Microcomputer, *J. Coatings Technology* **59**. (749) (1987) 31.
27. R.D. Peck, Computer-aided formulation, *Paint and Coatings Industry USA* **3** (2) (1987) 9.
28. S. Harshfield and J. Stopper, Microcomputers in paint industry formulation, *J. Water Borne Coatings* **6**. (1) (1983) 8.
29. J.T. Sheerin and M.R. Briber, Computers, chemists and costs, *J. Paint Tech.* **39** (515) (1967) 729.
30. H.G. Stephen, Is computer formulation possible? *Australian OCCA Proceedings and News* (June 1979) 9.
31. Product Formulation Expert System, Logica UK Ltd, Cambridge.
32. L. Cutrone, I. Melville and L.A. Simpson, Pigment thickener interactions in emulsion paints, *Polymers Paint Colour Journal* **177** (4187) (1987) 174.
33. Pigmentech Consulting, Jekyll Island USA.
34. PC Formulate, Rohm and Haas France SA, Valbonne, France.
35. D.G. Tsimboukis, Formula optimisation: feasibility study on methodology and techniques applicable. *Proceedings 4th International Conference on Organic Coatings Science and Technology*, Athens (1978) p. 606.
36 H.R. Touchim, Principles of design of experiments, *Paint, Oil and Colour Journal* (1963) pp. 83–86.
37 G. Buzzi Ferraris and E. Ranzi, Programmazione della sperimentazione industriale, *Italcartografica*, Italy (1968).
38. D.J. Rutherford and L.A. Simpson, The use of flocculation gradient monitor for quantifying pigment dispersion in wet and dry paint films, *Proceedings of 16th FATIPEC Congress*, Liège **3** (1982) 255.
39. L.A. Simpson, A method for assessing pigment dispersion in coatings using microwave exited oxygen. Tioxide Group PLC report number D9002GC (1983).
40. G.E.P. Box, W.G. Hunter and J.S. Hunter, *Statistics for Experiment.* John Wiley & Sons Inc, New York (1978), Ch. 10, p. 323.
41. Ibid. Ch. 6, p. 170.
42. J.W.H. Van den Haak, Computer assisted paint research, *Polymer Paint Colour Journal* **179**. (4243) (1988) 530.
43. R.R. Kuhn, Statistical design in formulating high solids coatings, *Modern Paints and Coatings* (August 1979). p. 32.
44. H. Grinsfelder, Experimental design revisited, reviewed and revised, *Resin Review* **20**. (2) (1970) 25.
45. M. Broder, I.P. Kordomenos and M.D. Thomson, A statistically designed experiment for the study of a silver automative basecoat, *J. Coatings Tech.* **60**. (766) (1988) 27.
46. J.C. Leveque, and A. Organeaux, L'utilisation des methodes statistiques dans la mise en oeuvre des pigments. Double liaison, *Chimie des Peintures*, (316–317) (1982) 30.
47. R.L. Plackett, and J.P. Burman, The design of optimum multifactorial experiments, *Biometrika*, **33** (1946) 305.
48. G.E.P. Box and J.P. Behnken, Some new three level designs for the study of quantitative variables, *Technometrica* **2** (1960) 455.
49. C.L. Vance, Statistical determination of numeric colour tolerances, *Modern Paint and Coatings*, (Nov 1983) p. 49.
50. P.J. Claringbold, Use of simplex design in the study of the joint action of related hormones *Biometrics* **11** (2) (1955) 184.
51. H. Scheffé, Experiments with mixtures, *J. R. Statist. Soc.* Series B. **20** (2) (1958) 344.
52. H. Scheffé, Simplex-centroid designs for experiments with mixtures, *J. R. Statist. Soc.* Series B. **25**. (1963) (325).

53. W.O. Thompson and R.H. Myers, Response surface designs for experiments with mixtures, *Technometrics* **10**. (4) (1968) 739.
54. N.R. Draper and W.E. Lawrence, Mixture designs for three factors, *J. R. Statist. Soc.* Series B, **27**. (3) (1965) 450.
55. R.D. Snee, Design and analysis of mixture experiments, *J. Quality Technology* **3**. (4) (1971) 159.
56. B.L. Hare and L.P. Brown, Plotting response surface contours for three-component mixtures, *J. Quality Technology* **9**. (4) (1977) 193.
57. L. Cutrone, Coatings formulation using three-component contour plotting, Tioxide Group PLC report number D9198GC (1985).
58. L. Cutrone and D. Becherel, Method for formulating paints using three-variable diagrams and a personal computer, *Pitture e Vernici* **62**. (9) (1986) 65.
59. J.E. Hall and A.P. King, Coatings formulation using four-component mixture design experiments. Tioxide Group PLC report number D7116 (1988).
60. J.E. Glass, *The Influence of Associative Thickeners and Rheology on Coatings Performance*, North Dakota State University Press, Fargo, North Dakota (1983).
61. Montreal Society for Coatings Technology, Influence of rheology modifiers on the performance characteristics of latex paints, *J. Coatings Technology* **58**. (738) (1986) 65.
62. J.E. Glass, Thickeners in perspective, *Amer. Paint Coatings Journal*, (August, 1984) p. 45.
63. E.J. Schaller, Associative thickeners, *Surface Coatings Australia*, **22** (12) (1985) 6.
64. J.E. Hall, Titanium Chelates. A novel approach to rheological control in latex paints, *J Coatings Tech* **59**. (749) (1987) 51.
65. B. Clementson, Titanium chelates. Tioxide International Ltd publication LP118 (1988).
66. R.D. Gorman, Simplex lattice designs for multicomponent systems, *Technometrics*. **4**. (4) (1967) 463.
67. R.D. Snee, Experimenting with mixtures, *Chemtech*. **9**. (1979) 702.
68. J.A. Cornell, *Experiments With Mixtures*, John Wiley & Sons. Inc. (1981).
69. K.K. Hesler, J.R. Lofstrom and J.A. Greenwald, The simplex coatings research, *J. Coating Techn.* **56** (713) (1984) 53.
70. G.P. Bierwagen, Computer prediction of latex paint properties, *Proceedings of 16th FATIPEC Congress*, Liège. **2** (1982) 13.
71. L. Cutrone, L.A. Simpson and I. Melville, Pigment thickener interaction in emulsion paints, Tioxide Group PLC report number D9209GC (1986).
72. F.B. Stieg, Effect of extender on crowding of titanium pigment, *J. Coatings Techn.* **61**. (778) (1989) 67.
73. J. Temperley, M.J. Westwood, M.R. Hornby and L.A. Simpson, A mathematical model for predicting the state of pigment dispersion in a paint film, *Proceeding of 20th FATIPEC Congress*. Nice (1990).

7 Design for bulk manufacture

IKE BERG

7.1 Introduction

7.1.1 *The basic problem*

The dispersion of particulate solids in a liquid medium has been practised for many years in a variety of disciplines. Consistency in quality has been maintained by the simple expedient of starting with a highly concentrated mix to ensure that the liquid medium is equally distributed around the particles. As a corollary it may be inferred that the particles should be wetted approximately to an equal degree. A slow reduction of total solids with more medium and, if necessary, other materials enables a satisfactory final dispersion to be obtained, provided that there has been sufficient agitation during the operation.

All processes, even the simple procedure above, create many problems when scaling up for bulk manufacture. These are especially exemplified in the paint industry with its numerous particulate solids (the pigments) and the wide variety of liquid media available. The precautions adopted and the procedures involved in the bulk manufacture of paints given in the following pages are in line with previous publications by the author, distributed in collated form with the 23 February 1983 issue of *Polymers, Paint and Colour Journal*, to whom acknowledgement is made for permission to republish in amended form.

7.1.2 *Conditions for dispersion*

Dispersion is achieved basically by a mixing process within which two types of flow can readily be recognised:

1. Turbulent flow which enables massive interchange of material on the macro scale between various locations.
2. Laminar flow, which occurs essentially in layers, involves material interchange between adjacent layers at the micro (virtually molecular) level. Laminar flow is therefore the requisite for the shearing action necessary to disperse particulate solids from the initial agglomerates to smaller clusters and ultimately to primary particles.

The transition between laminar flow and turbulence is not easily defined and it is quite usual with some machines for both types to exist in the same tank at different (and sometimes changing) locations. However, as the shearing action is a function of laminar flow, it is obvious that the greater the degree of laminar flow in a system the more efficient the dispersion. The higher the effective viscosity, the greater is the degree of laminar flow, which becomes more effective when the distance between the agitating member (ball, bead, roll or impeller) and the point of application is reduced. Increased temperature as a result of internal friction will reduce viscosity and loss of efficiency may occur unless the smaller particles released during the process absorb more liquid. The latter effect may be greater than the loss of viscosity due to heating and liquid may have to be added to allow the action to proceed. This situation generally applies with organic pigments and other high oil absorption materials.

7.1.3 *Particulate solids*

The pigments used in the paint and printing ink industries are legion. Apart from the obvious general mass properties such as relative density, colour, reactivity and resistance, the specific properties of the primary particles are significant. These include size, shape, cohesiveness and general surface properties which can be markedly affected by adsorbed impurities.

The shapes of the particles can often affect properties of prepared products, of which they form a part, and therefore their study can be crucial. However, when comparing particle sizes it is much simpler to consider the particles as spheres – of equal volume to the particles. Thus only one figure (the diameter of the spheres) need be used to compare the relative sizes of the particles of the different pigments.

Figure 7.1 illustrates the wide variety of particle sizes of some of the pigments used. It is obvious that some pigments have particle sizes 100–200 times (or more) the particle sizes of other pigments. It will not be surprising if problems arise as a result of attempting to disperse mixtures of particles with such large variations in size. Considering the basic dimensions, the ratios are as follows:

Ratio of particle diameters $1:10^2$.
Ratio of particle surface areas $1:10^4$.
Ratio of particle volumes $1:10^6$.
Ratio of numbers of particles to make up same total volume of material $10^6:1$.

It follows therefore that:

(a) When small quantities of the pigment with small particles are milled with large quantities of the pigment with large particles, precautions must be

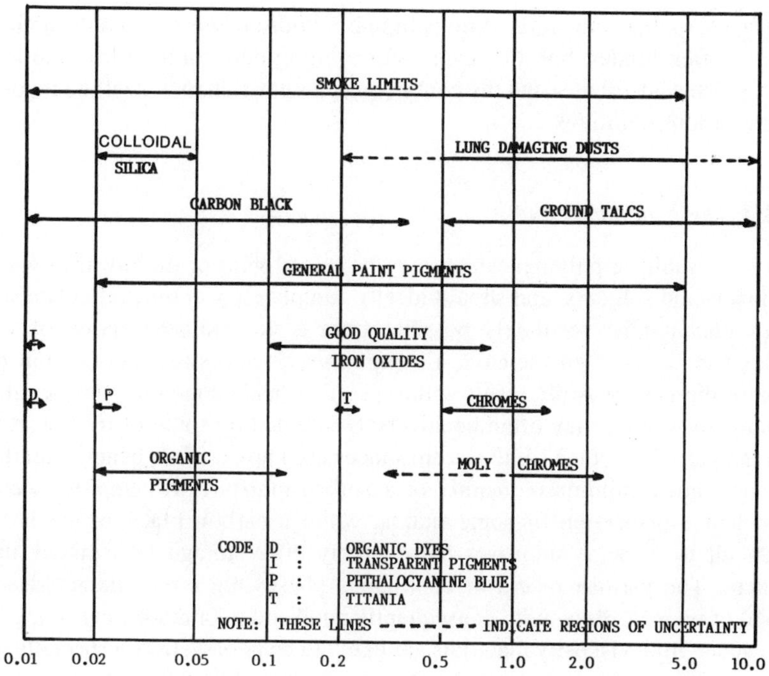

Figure 7.1 Particle size range of various pigments (in μm).

taken to avoid the situation in which the small particles are shielded within the larger particle interstices and are not subjected to efficient dispersion.

(b) When equal volumes of two pigments of such dissimilar particle sizes are milled together, the requirement of the smaller particle size pigment for adequate wetting (i.e. replacing the impurities adsorbed on the particles' surfaces by the continuous phase) will be so great compared with that of the other pigment that the latter can be swamped if care is not taken.

The influence of the pigment or other particulate solid may vary from quite a comparatively minor role, such as that for colouring only, to fairly important parts such as:

- protection from deterioration as a result of exposure, e.g. to ultra-violet, infrared and visible radiation;
- protection from corrosion;
- modification of rheology for specific application properties.

Whatever role the particulate solid is designed to play, however, the degree or efficiency of dispersion (and as a corollary, the degree of efficiency of wetting) will determine in no small way the success of the exercise.

Storage may have a profound effect on wetting (and consequently dispersion) because the particulate solids tend to adsorb various 'impurities' on the surfaces, such as moisture, various liquids, solids and gases from the ambient. These often hinder, but may even help, wetting depending on the characteristics of the adsorbates and the continuous phase which may itself be modified with various additives.

7.1.4 *The continuous phase*

This is usually a polymer which may be mixed with or dissolved in various adjuncts and solvents, and should ideally completely wet the particulate solid. This may not be absolutely possible from a stoichiometric point of view. Therefore, as is often the case, a compromise is necessary and the aim is to obtain the best possible result within practical and economic considerations. Although wetting may often be adversely affected by some of the adsorbates on the particle surface some circumstances are known when beneficial effects occur. Thus a minimum quantity of adsorbed moisture will improve wetting and hence dispersion of some titania, while a carbon black which is very difficult to disperse improves considerably after storage in a mixed atmosphere. The various media or continuous phases are often characterised as good or poor 'wetters' with consequently good and poor dispersing properties. However, high viscosity materials are likely to be poor wetters, especially with low shear stress machines. Therefore reducing a portion with solvent, as permitted by the formulation, favours the dispersion process, the remaining high viscosity fraction being added later. Another device to improve dispersion consists in adding specific surfactants. However, some of these can act both as dispersants and also as flocculants. Therefore additives should be avoided unless absolutely necessary.

7.2 Stages of bulk manufacture

7.2.1 *Introduction*

The object of a dispersion process is to obtain a stable base of the particulate solid in the vehicle. Unfortunately nothing is really stable and as soon as the mechanical forces used to produce the dispersion are stopped, there is a tendency for the particles to meet, coalesce into larger masses, finally to settle or flocculate. Therefore, in order to prepare a base of the required stability it is necessary to consider the following factors:

Pigment concentration. The higher the pigment content the more likely are the particles to meet and coalesce as a result of internal motion.

Degree of dispersion. The better the dispersion, i.e. the nearer the pigment has been milled to its primary particles and fully wetted, the longer will be the time taken for coalescence to sizes that would be detrimental to the final product.

Non-volatiles of the vehicle. For stability the particles must be adequately wetted by the polymer vehicle non-volatiles which also must form a barrier between the particles.

Viscosity. The higher the viscosity or structure of the material the greater the resistance to internal motion with consequent improvement of stability.

Within the overall dispersion process there are a number of conflicting requirements.

To maximise the opportunity for wetting of the pigment a medium having low intrinsic viscosity, i.e. a low solids vehicle, must be used with just sufficient resin solids to give a coating around each dispersed particle.

To maximise shearing forces during the dispersion stage, the effective viscosity of the mix needs to be as high as possible, depending on the equipment being used.

Once dispersion is complete, the overall pigment content needs to be reduced, and the resin solids increased to a point where the completed base has maximum resistance to flocculation and settlement.

This conflict in requirements is normally overcome by breaking the process down into several stages, i.e.

- The dispersion stage.
- The let down stage (stabilisation).
- The completion stage.

7.2.2 *Stable base formulation*

The above factors, in conjuction with experience, enable a base to be formulated such that it will be reasonably stable for a given time, depending on manufacturing procedures. Care must be taken because instability starts at the micro level and proceeds with increasing momentum. Furthermore only a very generalised formulation is possible, because of the variety of pigments and media available, and it should therefore be treated as a guide, based on practical experience.

Table 7.1 indicates the basic requirements of a stable base. The quantities given are based on oil absorption figures which refer to the quantity of linseed oil (or any other acceptable binder) which just binds 100 g of the pigment. It is conceivable that some formulations will not fit in perfectly with the recommendations, in which case experience will probably indicate the next best fit. It is emphasised that this table indicates the formulation of the stable base, i.e. that with which the dispersion process is completed. The formulation with which the process should start is discussed in the next section and indicated in Table 7.2.

Table 7.1 Stable base formulation

Pigments			
Oil absorption (g/100 g pigment)	Examples of types	Maximum pigment concentration %	Minimum non-volatiles of liquids %
Low (10–20)	Titanium dioxide, whiting chromes, natural oxides	60–70	30–35
Medium (20–40)	Synthetic oxides, china clay	30–50	35–40
High (40–60)	Organic pigments	15–30	40–45
Very high (>60)	High surface area pigments and extenders	10	45 <

Table 7.2 Initial dispersion (milling) formulation

		Vehicle non-volatiles (%) for machines	
Pigment oil absorption	Pigment concentration (%)	High shear rate (low shear stress)	Low shear rate (high shear stress)
Low	75–85	0–10	25–30
Medium	50–75	10–20	30–35
High	25–50	20–30	35–40
Very high	< 25	30 <	40 <

7.2.3 *Initial dispersion formulation*

The formulation with which the dispersion process is started must ensure that laminar flow is accomplished by high viscosity or consistency, preferably induced by a high pigmentation. It would be unrealistic however to indicate the viscosity required because:

1. It would rarely be Newtonian, unless at a very high temperature, and representation for repeatability would be virtually impossible.
2. At normal temperatures the viscosity prior to milling usually has a yield point, while the real relevance of the mill base consistency relates to the actual value(s) during the process.
3. The consistency changes during the process:
   (a) heat evolved has a profound effect;
   (b) inorganic pigments may result in a reduction of viscosity by the release of liquids held within the interstices of the larger clusters;
   (c) organic pigments tend to increase viscosity because the increase in surface area of the fine pigments requires more liquid for wetting.
4. Different batches of raw material may have slight variations in particle size distribution which generally would be perfectly acceptable in the finished product. The viscosity figures for the dispersion bases may, however, be sufficiently different to make control by viscosity unrealistic. The

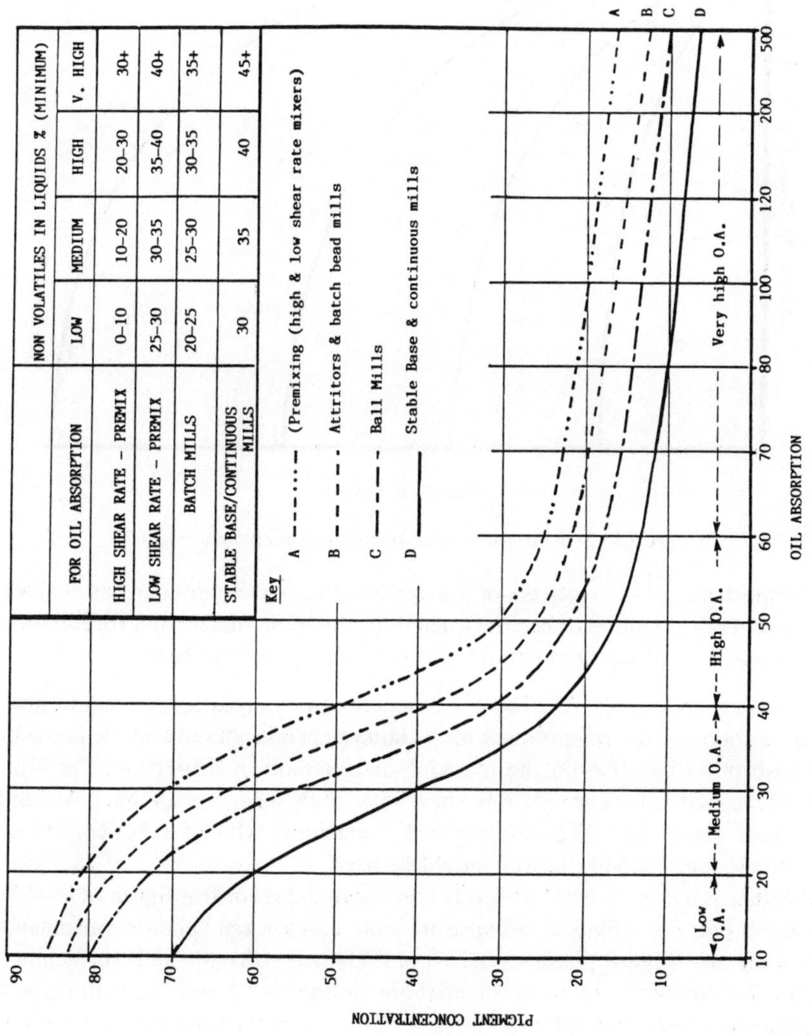

FOR OIL ABSORPTION	LOW	MEDIUM	HIGH	V. HIGH
	NON VOLATILES IN LIQUIDS % (MINIMUM)			
HIGH SHEAR RATE - PREMIX	0–10	10–20	20–30	30+
LOW SHEAR RATE - PREMIX	25–30	30–35	35–40	40+
BATCH MILLS	20–25	25–30	30–35	35+
STABLE BASE/CONTINUOUS MILLS	30	35	40	45+

Figure 7.2

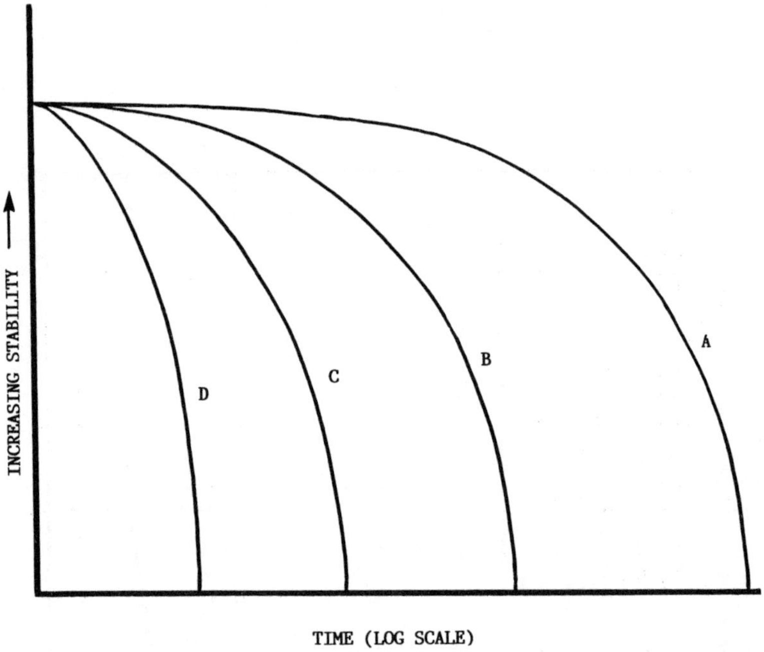

Figure 7.3 Stability curves after stabilisation at different stages.

importance of viscosity is not decried but practical experience has enabled a generalised guide to be constructed for the initial dispersion formulation (Table 7.2) which compensates for the rheology of the base.

As distinct from Table 7.1, Table 7.2 denotes two ranges for the liquid non-volatiles because the requirement of the latter depends not only on the pigment oil absorption but also on the machine and method of dispersion. For simplicity the two ranges are for low shear rate (high shear stress) machines and low shear stress (high shear rate) types. For machines which fall between these two types, intermediate figures should be used.

Another point to be emphasised is that Table 7.2 records a figure of 0–10% for the liquid non-volatiles because in some cases when specific surfactants are used the liquid phase consists entirely of solvents and surfactants. Figure 7.2 combines the recommendations of Tables 7.1 and 7.2 with respect to pigment concentration for the different types of milling machines specified, together with the stable base formulation which, as will be covered later, is also the milling formulation for continuous mills.

7.2.4 *Let-down stages*

It has been emphasised that (unless a stable base formulation is being milled as advised for continuous mills) once a base has been milled to the required dispersion, it must be stabilised. If the milled base is discharged without being stabilised, deterioration will develop at an ever increasing rate. Addition of materials to obtain a stable formulation always results in further flocculation although the increase in medium will tend to reduce the rate. The longer the time taken, after discharge, before a milled base is stabilised the shorter will be the useful life of the material. This is shown in Figure 7.3, which indicates the loss of stability with time when different procedures are involved, as follows:

Curve A. The material is stabilised in the mill. This is obviously the approach to adopt whenever possible because flocculation occurring as a result of adding the required stabilising components will be minimised provided the rate of addition is controlled and some remilling takes place. This will be discussed for the specific mills.

Curve B. The stabilising material is added immediately after discharge. However short the interval between discharge and stabilisation there is a pronounced reduction in resultant stability or shelf life of the product.

Curve C. Stabilising is effected some time after discharge. The impact on stability is significant.

Curve D. No attempt at stabilisation will result in poor shelf life and this procedure should be avoided as far as possible. It may be necessary in very specific highly pigmented protective compositions, in which case they should be applied as soon as possible after manufacture.

The curves are of a general shape and sequence for all manner of products but the actual time intervals will depend on the specific products.

7.2.5 *Completion*

The stable base is finally treated with various materials, bases or blended with other finished products to produce the required end product. Although the base and the additives are stable it must be recognised that some deleterious effect may be caused. This is usually very small because all materials to be mixed should be stabilised before addition. However, precautions are still necessary and all additions should be made while mixing as efficiently as possible.

7.3 Types of dispersion equipment

7.3.1 *Introduction*

The basic principles embodying the most efficient dispersion techniques are constant irrespective of the type of equipment employed. The application of the principles, however, must show some variation because of the differences inherent in the operation of the varied dispersion plant, as will become apparent when the latter are considered in detail.

Dispersion equipment may be subdivided into two main classes:

1. Batch mills.
2. Continuous mills.

Each class is capable of further sub-division, while it must be stated that a continuous process could at some given stage be fed with materials which may have been prepared by a batch process. This is noted so that the fact that a 'continuous' mill is being used does not necessarily mean that the whole process is continuous.

The major mills representative of the two main classes, are batch mills (ball mills, attritors, batch bead mills, mixers); and continuous mills (bead/sand mills, roll mills). All mills must be earthed satisfactorily and precautions taken to avoid static in the charge. One method is to ensure polarity in the milling base.

7.3.2 *Ball mills*

7.3.2.1 *General.* These mills are horizontally-mounted cylinders partially filled with milling media. Rotation round the axis is designed to cause the media to be raised to a given height and then allowed to cascade under the influence of gravity, to the bottom of the mill. Mill base in the interstices is therefore subjected to a shearing action, provided that laminar flow appertains, thus providing the dispersing action.

'Steel' ball mills, designed to cope with metallic milling media, have a metal liner and the larger types are water cooled. 'Pebble' or 'porcelain' ball mills have ceramic type liners and cannot readily be water cooled. The milling media for these latter mills are non-metallic.

Ball mills have been the work-horse of the paint and other industries for many years and, because over this period of time reasonable results have usually been obtained without too much worry, a certain laxity in procedure has developed. This has led to unsatisfactory results when dealing with the more sophisticated materials of recent years, although partially the trouble has been the mis-use of the mills. Nevertheless many ball mills are still in operation and continue to do sterling work.

7.3.2.2 *Speed of rotation.* The rotational speed must ensure sufficient cascading of the milling media. However, the concept of having a variable speed so that it may be varied with different materials being processed is quite unrealistic. A practical approach to the subject is vital, although much play has been made on the idea of insisting on the optimum speed being a specific fraction of the critical speed. However, the variation likely, even during the manufacture of one item, necessitates the compilation of a table (Table 7.3) indicating the range of speeds most likely to prove effective. The dispersion practitioner can then determine whether over a period a slight change in speed is advantageous.

Table 7.3 Ball mills : rotational speed

Ball mill internal diameter (feet)	Recomended speed of mill (rev/min)
1	48–52
2	32–36
3	24–28
4	20–24
5	18–22
6	16–20
7	14–18
8	12–16

7.3.2.3 *Milling media. Shape.* Three main shapes are available:

1. Spheres are the most logical type of media for ease of cascading within any circumference. Moreover, the smallest uniform interstices make spheres the most efficient shape for milling media.
2. Ovoids in the natural form of pebbles are the cheapest. There has been, however, a hypnotic compulsion for this shape and synthetic (composition) 'natural shape' ovoids have been available. With interstices that are quite irregular ovoids cannot be seriously regarded as efficient as spheres.
3. Cylinders (or rods) were originally conceived as joining up in the mill to form rollers along the whole length! This situation never occurs and the cylinders are completely randomised in position with totally irregular interstices. The same weight of cylinders would need less mill base than would spheres or ovoids in each batch operation. Cylinders must therefore be considered less efficient than ovoids or spheres.

Size. The smaller the milling media the more efficient is the dispersion. Figure 7.4 illustrates the results obtained by milling the identical mixture in (a) a lab ball mill with 12 mm balls, and (b) a lab bead mill using 1.2 mm beads. Although the speed of milling is vastly different (nearly a week for the ball mill; less than an hour for the bead mill), it is obvious the colour development in the ball mill will never equal that in the bead mill. Milling media should therefore be as small as possible, but note:

- The smaller the media the more buoyant are the members (because of large surface area) within the paste, reducing the efficiency of cascading.
- The smaller the media the more paste is retained on the surface, i.e. it becomes more difficult to empty and clean the mill.
- The media must be big enough to be retained in the ball mill when emptied.

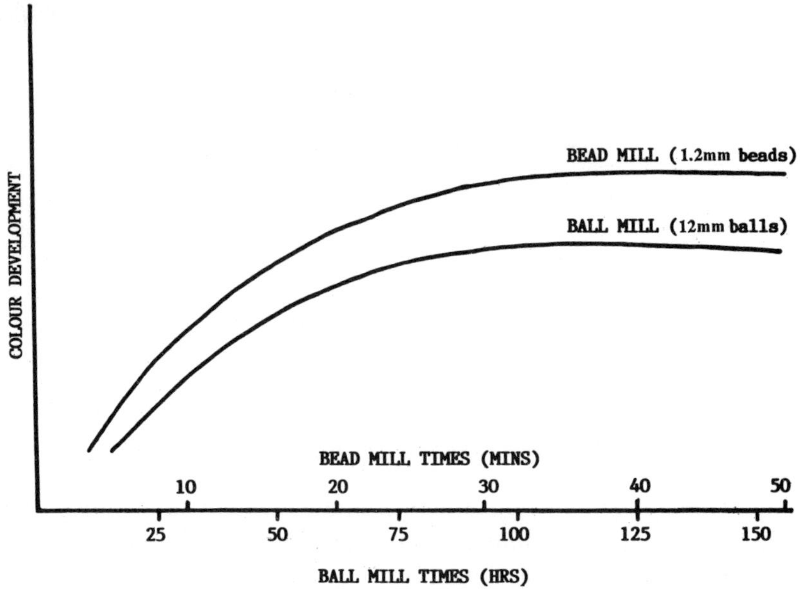

Figure 7.4 Effect of milling media size in colour development.

For the best ball mill work media exceeding 15 mm in diameter should be avoided.

Relative density. Other factors being equal, the higher the density of the milling media the better is the degree of dispersion in a given time. However, the main advantage of higher density media is the fact that higher viscosities (and therefore higher pigment loading) are possible which definitely enhances the efficiency of dispersion. The density of the media is of course a function of the composition which is discussed below.

Composition. The composition of the milling media determines both the density and wear. It is therefore necessary to check details.

Metallic milling media should be through-hardened by heat. It may not be wise to obtain 'work hardening' media which may wear too much before work hardening occurs. Metallic media that are not through-hardened will wear rapidly into unsatisfactory shapes. Apart from through-hardened ball-bearings, other media prepared from high nickel or high chrome content steels have been satisfactory.

Non-metallic milling media, which were the one-time cheapest, i.e. the natural ovoid pebbles, are now almost unobtainable and therefore synthetic types are virtually absolutely necessary. These may be based on steatite, other ceramic types (including porcelain) and alumina. Some of these are described as high density but the description should not be applied unless the relative density is 3.5 minimum, i.e. about 50% in excess of that for the general pebble/ceramic media.

Laboratory or works scale checks are necessary to check wear, but it can be categorically stated that media prepared by a sintering process should be avoided. These usually exhibit a high rate of wear while contaminating the pigment base with an unwanted extender. Fused media are necessary and even here the variation is such that tests are still recommended. It should be noted that cylinders (or rods) having circular but sharp edges can be rapidly eroded in those areas.

Mill charge. The mill should be charged with one size of milling media. The practice of having two or more sizes has no real foundation. In fact during the milling process the media often partially classify; if they do not classify but remain well mixed, the interstices are randomised and become larger, reducing the process efficiency.

Non-metallic media, whether ovoids or spheres, should be charged to approximately half way. However, if using very high density materials, care must be taken to ensure that the liner of the mill will accommodate the pounding due to the high weight. If not, it may be wise to reduce the actual charge of media accordingly, as is often the case with metallic media. Non-metallic media may often be loaded with advantage into a steel-lined mill for cooling effects.

Metallic media are invariably loaded into a steel-lined mill, never into a 'porcelain' mill. Provided the mill and motor can accommodate the weight of metal, the steel-lined mills are ideally loaded with the metallic media up to half way as for the non-metallic media. However, some mills may be worn, under-powered or not designed for the full load. In such cases the more usual loading is 33⅓% apparent volume of the mill.

Mills which have smooth sides (steel mills) or which have very high density media, or which are charged only partially with media (rather than full 50% charge) should be provided with 'lifter bars' to help raise the media to the cascading position. This is invariably the case with steel mills (except the smallest and lab mills) and sometimes the position with the larger porcelain mills.

In order to calculate the charge by weight when the relative density is known the following may be used:

- spheres will have a void volume of about 35% of total apparent volume;
- ovoids will have a void volume of about 40% of total apparent volume;
- cylinders will have a void volume of about 30% of total apparent volume.

7.3.2.4 *Milling capacity.* For the most efficient milling procedure the volume of the pigment mill base to be processed should initially be equal to the effective voids of the milling media, while the mill is operating. This volume would obviously be greater than the voids of the media at rest and, expressed as a percentage of the total apparent volume of media, would be of the order of 40% for spheres, 45% for ovoids and 35% for cylinders or rods.

The total apparent volume of the milling media is, of course, the actual volume of the solid media and the voids. Therefore, if the mill is charged with milling media to an apparent value of 50%, the volume of the pigment mill base expressed as a percentage of the total mill volume would be 20% for spheres, 22.5% for ovoids and 17.5% for cylinders. If the apparent volume of the milling media is not 50% then the volume of the pigment mill base must be changed accordingly. Overcharging the mill with pigment mill base should be avoided at all costs because efficiency of milling is thus drastically reduced. Extra milling time will not compensate for the reduced milling efficiency.

7.3.3 *Attritors*

7.3.3.1 *General.* These machines are vertically standing cylinders, usually filled to approximately three quarters of their apparent volume with milling media. The cylinder, usually of hardened steel, does not rotate. The milling media, however, which may be metallic or non-metallic, are moved around by means of rods attached to a central rotating shaft, driven by an electric motor.

The original (S-type) attritors were charged at the top and emptied at the bottom, the milling media being retained by a screen. However, the larger machines are usually designed with a circulating system which avoids poor millings at otherwise dead or sluggish spots and provides an alternative discharge point.

The Q-type attritor is basically a modified S-type which is connected to a high-speed mixer/disperser (which will be discussed at a later stage) such that the mix, i.e. the pigment mill base, is circulated continuously between the attritor and the disperser, by means of a pump and other connections, while both machines are operating.

The high-speed mixer/disperser is of a greater capacity than the attritor, allowing a batch of material up to ten times that normally made by the same size S model. The claim for the Q model is that the total time for manufacture (e.g. for making the batch ten times that of the usual S size) is much less than the manufacture of the (ten) single batches, sometimes saving up to 40% of the time.

There is some substance in this claim, but the time saved varies considerably, depending on.

- the formulation;
- the size of the original S machines, some of which are modified differently to make them more efficient, e.g. with more media and the addition of a grinding surface-cum-strainer.

Nevertheless it is an interesting machine for larger batches, especially for bulk primer and surface manufacture.

7.3.3.2 *Milling media.* The original UK suppliers of the attritor (Torrance and Sons) had advocated a mixture of $\frac{1}{4}$ inch and $\frac{5}{16}$ inch diameter spheres as milling media. Experience has shown that this is not ideal and only one size, preferably $\frac{1}{4}$ inch, should normally be used. Mixtures of sizes, if they do not result in layering, will certainly increase the voids size and therefore reduce efficiency. However, the dispersion practitioner will make his or her own decisions depending on material to be dispersed, such as the use of $\frac{1}{8}$ inch diameter metallic spheres when dry grinding various metallic oxides.

It would obviously be unrealistic to have shapes other than spheres for the milling media, the composition of which may be similar to that for ball mills. Sintered media are not advisable. Metallic media may be used for dark colours and non-metallic for clean colours. However, because the usual attritor is metallic the colour may not be quite up to the standard obtained in porcelain ball mills. Therefore for absolutely clean results it may be necessary to resort to a machine lined with non-metallic (e.g. ceramic) material. This would introduce problems of cooling more especially in the larger models.

7.3.3.3 *Milling capacity.* The pigment mill base should not exceed in volume the voids of the milling media in motion, i.e. approximately 40% of the total apparent volume of the milling media (spheres and voids). Overloading the mill, that is exceeding this volume of pigment base, will necessitate longer operation times without necessarily achieving the required result.

7.3.3.4 *Comparisons with ball mills.* It has been claimed that the attritor is about ten times as fast as a ball mill. Broadly speaking this is correct. The reasons for this faster milling can be deduced as follows:

(a) The attritor is actuated by means of an electric motor which rotates the milling media. The media in a ball mill is caused to cascade by means of gravity. If the viscosity tends to reduce the cascading efficiency in the ball mill there will be less efficient milling. In the attritor, however, a higher viscosity (and therefore higher pigmentation and more efficient dispersion) is tolerated.
(b) The milling media in an attritor are smaller than in the conventional ball mill. Reference has been made previously to the improved milling associated with the reduction in size of the milling media and exactly the same argument holds. Smaller media mean smaller interstices with greater control for laminar flow.
(c) When operating the attritor most of the separate members of the milling media are in motion. With ball mills, however, a large proportion of the media at any one time may be quite motionless or be moving very slightly until they reach the position in the mill when they start to cascade. The larger the mill, the larger the volume of media affected by the conditions noted above.

It must be remembered, however, when comparing times of milling, that the ball mill is usually allowed to operate unattended overnight because it is an enclosed type of machine. The attritor, if operation for fractions of a day is necessary, must be attended for the full time unless arrangements are made for an extended day or left running with full safety devices. This may not be popular, especially these days, and the time saved may not be worth all the safety and other factors. Therefore if it is realistic to assume that S type attritors are likely to run for the eight-hour customary day, then the factor of ten times faster than the ball mill must be reduced to three or four times faster. Of course, if an extended day is run the factor will be increased.

Most materials that are processed on ball mills are readily made on attritors, those requiring steel ball mills usually being processed with metallic media as would be expected. There are, however, some examples where transfer from ball mill to attritor manufacture may present some difficulty:

- Materials that are required to have an exceptional cleanliness and which can only be obtained from a pebble or porcelain ball mill with non-metallic media. Such products would require the attritor to be wholly constructed or lined with a ceramic type of material. This may be costly or difficult to operate.
- Products that have specific organic extenders which are hammered by the balls in a ball mill to give a specific fairly repeatable required pattern. Although this is strictly not a dispersion problem, it may present some difficulties to transfer from a ball mill to an attritor.

7.3.4 *Batch bead mills*

7.3.4.1 *General.* These machines in their simplest form are like the attritor, being vertical cylinders in which the media are rotated, the shell remaining fixed. The original Berger-designed mill grew from a simple design and some relevant details are as follows.

Milling media. These were originally glass beads 1.2–1.5 mm in diameter. The quantity of milling media was approximately 0.5 kg/litre total empty volume of the mill. Thus the apparent volume of milling media was less than 30% of the total empty capacity of the mill shell.

Power. For adequate power a minimum of 1 kW for every 10 kg milling media is suggested.

Disc. The size of the disc positioned to aid discharge must be adequate to cause all the milling media to be sufficiently well rotated when the machine is in operation. Although a figure of 80% of the mill diameter is suggested to be a reasonable diameter for the disc, it must be remembered that scaling up (or down) changes the distance from the disc edge to the tank (shell) side. If this distance is too large, slip may occur which will reduce efficiency. The smallest laboratory-type machine will operate successfully with plain discs. Larger machines have either modified types of impeller discs or have holes in the discs to enable the milling media to be actuated easily. The peripheral speed of the disc is of the order of 750 m/min for general purposes. It should be noted, however, that if for some reason higher speeds are desirable, there would be a limitation imposed upon the size of glass milling media, e.g. glass beads of 3 mm in diameter would tend to shatter at speeds exceeding the figure noted above. The disc should be manufactured from hardened or chromium-plated steel.

Apart from the modified types supplied by Torrance, it must be mentioned that other types of batch bead mills have been available that are different from the simple type in many respects, but mainly from an operational point of view, in having a method of discharge based on centrifugal action. These mills, of which the Centrimill (OBS Machines) and Monomill (August) are examples, are treated in the usual way for the best procedure, i.e. the pigment mill base must be equivalent to the void volume when operating (i.e. about 40% of the total apparent volume of the media and voids). With a lower peripheral speed some suppliers advocate the use of 3 mm beads for their models.

7.3.4.2 *Method of operation.* The batch bead mill, with smaller milling media and faster rotation, disperses at a faster rate than the attritor. However, if very

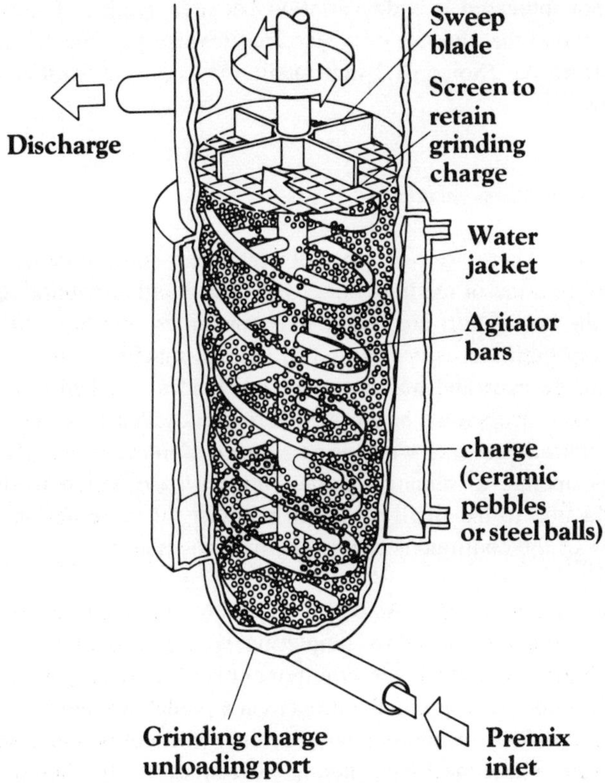

Figure 7.5 Attrition mill.

coarse pigments or extenders are used, the glass milling media may wear quite rapidly. There is of course the possibility of using different composition beads, but here again caution is advised and sintered media should not be considered.

If the simple type of batch bead mill is used, there is a tendency for some inadequately milled paste to accumulate in the region between the screen and discharge valve. Therefore, when the milling is virtually completed, some of the milled base is discharged and recharged at the top to wash out the poorly milled base back into the mill for adequate milling while a similar procedure is adopted during the early stages of stabilisation. Alternatively, a circulation system can be used.

This procedure is, however, unnecessary in the modification introduced by Torrance (who supply this type of machine), by having a precision-built valve fitting into the screen-discharge region which does not allow build-up to occur.

A final point to bear in mind is when charging or changing media. Some suppliers have been known to switch to different manufacturers of glass, with the result that sometimes leaded glass and at other times non-leaded glass is available. If charging is by weight the relative density should be carefully checked every time, to determine required volume. The rate of wear of glass media has not indicated a wide variation between grades of glass. It will, however, be found that the rate of wear varies inversely as the diameter of the media; therefore the choice of 2 mm media may be wise from an economic point of view.

7.3.5 *High shear stress mixers*

7.3.5.1 *General.* There are two types of high shear stress mixers, depending on the axis of rotation of the impellers, i.e. vertical and horizontal. Both types (especially the horizontal) are heavy duty machines, and are constructed to cope with very high viscosity bases. The vertical machines are usually operated with change pans and may have scraper blades attached to the rotating axis. If not, the process must be stopped periodically to scrape down the sides to allow an intimate mixing with the remainder. Cooling is usually provided. Both mixers operate to obtain a stable formula base which is suitable for relevant reduction to the finally finished paint or for presentation to another mill (usually of the continuous type) for final processing.

7.3.5.2 *Operating principles.* As with every other mill, the procedure adopted should be that which, if carried to completion, would result in the best possible dispersion obtainable, even if the requirement is of a lower quality. This may readily be obtained by stopping the process at a predetermined stage. By these means repeatability of results is assured: this would not be the case with any other procedure. These machines operate most successfully when the viscosity of the material being processed is at its maximum, therefore similar machines constructed from different materials may not have identical outputs; the more efficient output would be from the machine constructed from the more sturdy material capable of dealing with the higher viscosity. However, it must be emphasised that viscosity may vary during processing:

(a) Liquids trapped in the particle interstices may be released to reduce the viscosity, allowing more pigment to be added for greater efficiency.
(b) Heat is developed as a result of internal friction. The water cooling does have some effect but some more pigment can still be added as in (a).
(c) Small particle pigments increase their total surface area with continued dispersion, reducing the available liquid by surface adsorption. Therefore more liquid may be necessary to avoid the machine seizing up. In every procedure it is necessary to reduce as far as possible with the liquid available before stopping work, otherwise overnight cooling may result in a 'solid' mass in the machine which may have to be dug out.

7.3.6 *Low shear stress mixers*

7.3.6.1 *General.* Numerous types of these are available under the descriptive name of 'high speed mixers'. The original type consists of a disc attached to a centrally rotating shaft in a round tank. The disc may have a serrated edge or various attachments in order to increase the motion of the liquids, and its speed may be fixed (two or three speed), variable over a given range or infinitely variable to a given speed. The shaft can be centrally placed or movable to an eccentric position and fixed or movable up or down. The tank can be fitted with an operating scraper and be fixed or mobile, with or without water cooling.

7.3.6.2 *Main functions.* As with other mixers, the main functions of the high speed mixer with respect to dispersion operations are two-fold:

(a) as a premixer, i.e. to prepare a mill base formulation for subsequent milling in another (usually continuous) mill;
(b) as a premixer/mill combined, i.e. to accept the raw materials and process them sufficiently well that they can be used without further milling for completion to the required product.

Once again it is emphasised that with respect to (a) and (b) above, whether the mill is being used as the final (or only) mill or whether it is being used to prepare material for another mill, the same procedure and care must be applied, otherwise the second mill may suffer undue wear or may even not produce the required quality. It cannot be repeated often enough that the premixing is generally the most important part of the whole dispersion process.

Other functions of the high-speed mixer include the following.

Mixing, when the turbulence is allowed full sway. Thus when dissolving resins in solvents, pigment chips into resin solutions would be apposite.

Heating small batches (usually in mobile tanks) for specific purposes:

- When colour matching certain products the colour is 'fixed', i.e. not flocculated when the temperature is maintained at 60°C or above for a given time.
- When using specific thickeners based on castor oil derivatives, they have to be incorporated at elevated temperatures, depending on the composi- tion of the product processed, for required results.

7.3.6.3 *Operating principles.* A general idea of the parameters involved can be appreciated by examining what has been accepted or promoted as average, e.g.

Impeller (disc) diameter	$= D$
Tank diameter	$= 2\frac{1}{2}D \pm D$
Depth of mix	$= 2D - 2\frac{1}{2}D$
Height of impeller above bottom	$= \frac{1}{2}D - 1D$
Peripheral speed of disc for dispersion	= 23 m/s

The above will now be discussed in detail:

Ratio of tank diameter to disc diameter. It will be recalled that starting with a laboratory mixer the ratio of 2½–1 tank-to-disc diameter means that the distance between the edge of the disc and the side of the tank is only an inch or two. When scaling up, the same defined distance becomes a foot or two. This means that the shearing action of the disc periphery is damped down considerably in the large plant compared with the laboratory plant. It is found that sometimes very little motion is observed round the sides of the large tank, whereas the same formulation gave more than adequate mobility in the laboratory. Therefore one of three changes may have to be made.

1. The formulation will have to be reduced in viscosity for factory processing. This will reduce the efficiency of premixing – at the very least it will increase the time required for premixing, or may not even result in a satisfactory premix, however long the operation.
2. The peripheral speed should be increased. This will in the first case prove very expensive because the power requirement varies as the square of the speed. Moreover, there would be very little improvement in overcoming the damping effect of the liquid.
3. The size of the disc relative to the tank should be increased. This procedure does make a profound difference even if the peripheral speed is constant. This factor was very evident when easily dispersible organic pigments were first introduced. The original trials on a laboratory mill were excellent but poor

results followed on a large scale. It was then agreed that the ratio of discs to tank should have been nearer 1–1½ than 1–2½ when transferred to production size.

From the above it is obvious that dispersion efficiency is dependent on the ratio of disc to tank and the greater this ratio the more efficient the operation. On the other hand, this procedure of increasing disc size is also expensive, firstly because the power is a function of the cube of the disc diameter, and secondly, if the tank is shrunk (rather than increasing disc size), smaller and maybe uneconomic size batches would result.

Before leaving this section it should be stated that when using easily dispersed pigment, such as certain titanium dioxides, a low ratio of disc to tank such as 1–3½ and even 1–nearly 4 has given good results.

Depth of mix. It is unlikely that the depth of mix should be any greater than that given. Indeed, if very high consistency mixes are processed it may be necessary to reduce that depth to maintain satisfactory movement for processing to critical requirements.

Height of impeller above bottom of tank. If the shaft can be raised and lowered, the ideal can be adjusted for each trial. On the other hand, if the whole is fixed, the position of the disc must be arranged so that it is not too high for the initial liquid charge and not too low for effective operation when fully charged. Some basic formulations to be processed should be studied before designing the plant.

Peripheral speed. Usually the higher the peripheral speed the better the dispersion because the greater turbulence thus induced can be treated with more pigment to reduce the turbulence by increasing viscosity, creating the conditions for laminar flow. However, the author has dispersed certain titanium dioxides, specifically designed to disperse easily, at a peripheral speed of 12 m/s in a tank about 3.75 times the disc diameter. This fact is mentioned because it is emphasised that nothing should be regarded as fixed absolutely – the principles should be known and used in a flexible manner.

7.3.7 *Other types of mixers*

7.3.7.1 *Combined heavy duty and high speed mixers.* A heavy duty mixing system, usually with a scraper device, is arranged to work with a higher speed mixer inserted at an angle to avoid fouling other moving parts. This system is quite useful when consistency presents a problem. For general purposes, however, the dispersion results must be compromised if the viscosity is to be satisfactory for the working of both systems simultaneously, unless, of course, the pigments have been treated for ease of dispersion.

7.3.7.2 *Rotor/stator mixers.* These are usually used in line. The rotor at speed passes the liquid through slits or holes in the stator and the shearing action is claimed to enable adequate dispersion. However, a lower viscosity than can be used with ordinary high speed mixers is necessary for this action and would therefore reduce efficiency, thus nullifying any major advantage claimed in dispersion efficiency. There are advantages which may be justly claimed in specific operations such as 'in-line' requirements.

7.3.7.3 *Rotating tank mixer.* This is an excellent variation which is very suitable for producing premixes especially for high consistency. The tank rotates on a central plinth and a fixed scraper blade is designed such that material from the side is scraped off and flung into the operating vicinity of the high-speed mixing disc which is eccentrically placed. However, the safety regulations require specific constraints on construction which makes this machine quite expensive.

7.3.8 *Continuous mills*

7.3.8.1 *General.* The continuous mills used in the paint industry are fundamentally of two types:

1. Roll mills – the shearing action occurs between pairs of rollers or between one roller and a flat bar.
2. Sand/bead mills – the shearing action occurs as a result of the agitation of small milling media (sand, glass or other synthetic types).

Although there are specialised mills of each type which can cope with very coarsely prepared mixes and act as comminuting mills, these are the exception rather than the rule. For most mills of the continuous variety in the paint industry, a thoroughly prepared premix as previously described is necessary for them to function satisfactorily. Here again it is stressed that the result obtained will be greatly dependent on the efficient (or otherwise) manner in which the premix has been prepared.

The main difference between the two types is in the consistency of the premix for the best results. The roll mill performs most satisfactorily with a premix of very high consistency. Sand/bead mill premixes are of a lower consistency because of the mill construction, but nevertheless should be as high as the particular model will allow.

Continous mills have usually been fed with premixes that mainly have been prepared batchwise so that adequate control of the quality of the premixes is possible. However, the continuous preparation of premixes would not be impossible with a number of in-line premixer blades, especially with the more easily dispersed materials.

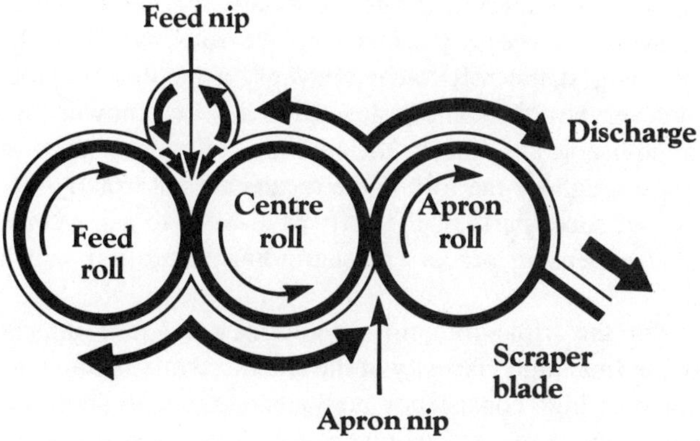

Figure 7.6 Triple roll mill.

On the other hand, as the discharge from these mills is stable, completion could be continuous. At present, however, batchwise working is the normal procedure except for some isolated instances.

7.3.8.2 *Multiroll mills.* The triple roll mill is the most widely used multiroll mill and the principles of operation can be readily appreciated from Figure 7.6. Other multiroll mills (e.g. five and seven roll) can be understood by extending the arguments for the triple roll mill.

The three rolls, feed, centre and apron, rotate in the directions indicated but at different speeds, e.g. if the rolls were 12 inches in diameter and 30 inches long the speed of the feed roll may well be 30–40 rpm. The centre roll in some of the older models could be rotating at 1.5 times that of the feed roll and the apron roll at 1.5 times that of the middle roll. However, the rate of production increases with higher roll speed ratios and the newer models tend to have a ratio of 2 and even 3.

The feed is through a 'hopper' formed by the feed and middle rolls with two end plates specially designed to ride on the rolls and prevent spillage out of the sides. The gaps between the rolls are maintained under hydraulic pressure at about 1–2 ml.

In operation, the very high viscosity of the stabilised pigment mill base combined with the different roll speeds ensures differential laminar flow of the paint entering the feed-middle roll nip. The close proximity of the rolls determines a very highly efficient shear stress action on the thin ribbon of material passing through. All the material eventually passes through both nips (where the action is similar) although a portion tends to be returned to the feed by residual pick-up of the rolls. A well designed and properly placed take-off apron can usually be nearly 100% efficient.

The triple roll mill is probably the most efficient dispersing machine available to the paint industry but certain precautions have to be observed.

Good training and supervision are necessary for the operation of these machines to avoid unnecessary damage to the rolls, which will wear even with the best care. If the rolls wear unevenly a number of problems arise which can be very troublesome unless the cause is known. To avoid any doubt it is advisable to check the milling efficiency at a number of points along the length of the roll. If the results are not exactly the same then it is obvious that some parts of the rolls have worn to a greater extent than others and the expensive process of regrinding the roll(s) will have to be considered.

Premixing for the triple roll mill operations must take into account the necessity for the final high viscosity of the stabilised mix. Although it has been possible to prepare high consistency products on the high-speed mixer, really high viscosity premixes would require not only heavy construction but high horsepower to cope with the forces necessary to move heavily bodied materials. The most suitable premixer for triple roll mills would be a heavy duty mixer of the high shear stress–low shear rate type. The combination of the latter type of mixer and the triple roll mill would provide a dispersion facility of the highest order – from the quality point of view – because each machine can be arranged to produce absolute laminar flow at highest viscosity.

7.3.8.3 *Single roll mill.* The principles of operating the single roll mill (or bar mill) is given diagrammatically in Figure 7.7. The stabilised premix in the hopper feed is dragged through the gap between the roll and the bar and is collected by the take-off apron. The bar which has two or more vanes – depending on the size of the mill – is held against the rotating roll by hydraulic pressure and thus serves a similar role as a roll on the triple roll mill. However, because the bar is static it is usually necessary (based on the feed composition) for the premix to be a little less viscous than that for the triple roll mill. Nevertheless it is an excellent mill with results almost up to triple roll mill standard and the principles for premixing will be virtually the same as given for triple roll mill manufacture. One major feature should be noted. The construction allows some paste through at the ends of the bar without being fully subject to the shearing action. This means that some poorly milled material may be collected. However, the take-off apron is so designed that the product collected is separated from the amount at each end which is likely to be poor. The end portions are therefore collected separately and returned to the hopper for further milling.

As with the triple roll mill, the single roll mill requires a well-trained operator and both the bars and rolls must be checked to make sure they are satisfactory. Regrinding the bar as well as the roll may be necessary at times.

For many of these reasons and because both mills are rather labour intensive, the triple and single roll mills are somewhat out of favour with paint manufacturers who are now most likely to be sold on the sand mills.

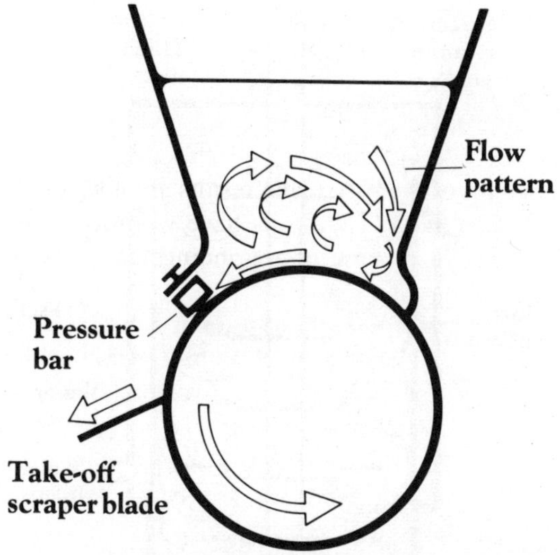

Figure 7.7 Single roll mill.

7.3.9 *Sand mills*

7.3.9.1 *General.* Continuous sand or bead mills are basically cylinders in which milling media such as sand, beads, etc. are actuated by means of disc type impellers while a premixed pigment (or other particulate solid) base in a resin, or similar, solution is passed through for processing. The base is discharged in a milled or dispersed state depending on the formulation and procedure (see Figure 7.8).

A number of different suppliers extolling their own design have made the choice of mill an interesting if difficult one. Whatever claims are made, however, some basic points apply, not only to the different suppliers' models but also to modern developments. Caution is necessary in assessing the changes in performance a modified design may cause but for all models it can be said that improved dispersion results from:

(a) A higher consistency of premix – provided that the machine is designed to cope with the high consistency;
(b) Smaller milling media – although lower limit is usually 0.8 mm;
(c) Greater quantity of milling media (i.e. increase in ratio of milling media to pigment base);
(d) Faster agitation of the milling media – this necessitates adequate horsepower and will affect temperature and pressure;
(e) Slower passage through the mill – also affecting temperature and pressure.

Other factors which may affect output rate and quality, especially with the later models are:

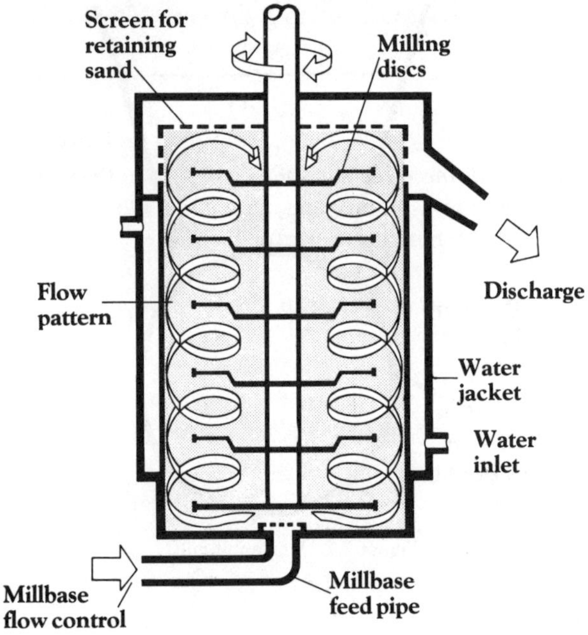

Figure 7.8 Sandmill.

- design of screen;
- design of the impellers;
- composition and size of milling media.

There are three main types of continuous sand mills: vertical open; vertical closed; horizontal.

There are other minor variations which include two or more mills in series but the method of operation for the best technical results follow logically from a study of the usual types. There are quite a number of different models of the above, although most suppliers at present tend to concentrate on the horizontal.

The differences in construction are mainly concerned with the following.

Discs. There is a wide variety of shapes of disc impellers in all three types. Many of the different shapes are claimed to be 'scientifically' designed although the author considers the main idea was to show they are different from others!

Seals. The seals are very important for the vertical closed and the horizontal types and performance should be critically considered.

Screens. The means of separation of the milled paint from the sand was a reasonably simple screen for the original vertical open mill. When the processed materials were shown to be continually building up on the outside of the screen the vertical closed mill was introduced to overcome this. This latter mill operated under pressure and the milled paste was thus pushed through the screen. At this stage a slit or dynamic separator was introduced by Drais and formed one of the methods used in horizontal mills, which use a variety of different screens depending on the makers' ideas.

7.3.9.2 *Comparison of the three sand mill types* All types of continuous mill will of necessity require a premix and, as has been previously stressed, the better the premix the better the result (or the easier it is to obtain the final result). All models similarly will of necessity require a stabilised premix, which in operation will pass through the mill and be subjected to a milling action by the media.

The simplest sand mill, i.e. the vertical open, usually performs quite adequately. The degree of milling may be controlled by reducing the rate of processing to improve the quality while increasing the rate where the premix is so good that milling occurs quite easily. It would be quite difficult, however, to cope with materials of high consistency or thixotropy.

The vertical closed mill can deal with higher viscosities and therefore basically must be regarded as a better mill.

The horizontal mill, however, with the possible very high milling media to mill paste potential, has the means of producing the best milling from any of the sand mill group. However, a number of these supplied by various manufacturers show differences in the design of discs (impellers), seals and screens and these will determine to a large extent some of the potential. If the screen is such that the area is fairly limited the pressure may build up quickly if fast pumping is used to increase the rate of production. Similarly the temperature may tend to rise. Thus the enormous benefits derived from the use of the horizontal mill must be tempered with a greater degree of control.

7.3.9.3 *Milling media.* The general principles previously noted for milling media hold but special considerations may apply as follows:

Shape. The media are subject to high speeds when the mill is in operation, therefore they must be, or approximate to, spheres.

Size. For different purposes different sizes may be used in the range (of diameters) 0.6–3.0 mm.

Composition. This can be either natural or synthetic.

Natural. The only natural milling media for sand mills is Ottawa sand, with a particle size of about 0.8 mm ± 0.2 mm and almost spherical. Its relative density is of the order of 2.6 and is generally sufficiently hard for pigment dispersion. New deliveries of this material should be recirculated in a resin solution through the mill and a screen until extraneous dust and jagged edges have been removed.

Synthetic (synthetic types include glass and ceramic). Glass has proved to be an excellent milling medium and can be obtained in various sizes. Precautions with respect to relative density are necessary because some suppliers just send stocks obtained from various places and are unaware whether lead is present in the composition. Wear is not so greatly different no matter what type of glass is used unless poorly shaped specimens are used in the first place. Steatite would be satisfactory if a suitable supply is obtained. Alumina would be suitable only if fused, not sintered.

Zirconia was recommended at one time. This oxide of zirconium was considered to be ideal because its relative density was around 5.6. However, the type available had been prepared by a sintering process which caused it to wear much more rapidly than would be acceptable and the processed material finished up with an unwanted extender. Moreover, the same medium within a month of use had caused considerable wear on the sand mill in which it was being investigated. Had the zirconia been prepared by a fusion process wear might not have been so pronounced both of the mill and media, while the high relative density could well have proved an advantage in spite of the large and variable sizes of the media.

Zirconium silicate prepared by a fusion process in the form of small beads has proved the most suitable medium yet. Whether it is a pure zirconium silicate or a physical fusion of the oxides of zirconium and silicon is not established and probably does not matter. However, with a relative density of 3.8 it has shown quite a number of advantages over glass with respect to milling efficiency:

- The higher relative density produces results faster for the same quality or a better quality for the same time of milling.
- There is less wear on the mill and the media wears far less than glass to the extent that although the medium may orginally be more expensive than glass, the reduced wear makes it effectively cheaper.

There is one drawback, however, which must be checked. Some poor batches of these zirconium silicate beads have shown a variable hollowness. In many batches the degree of hollowness has been slight, but where this has been excessive not only does this affect the relative density but many beads may actually break up. Therefore, although they are recommended, they must be checked for the hollowness described.

7.3.9.4 *Milling media charge*. The vertical open mills are the only ones which are recommended to be charged with sand. The quantity advised is best initially left to the supplier's ideas but can readily be adjusted by the dispersion technician. Usually the sand level is just above the upper plate but below the lower part of the screen. This loading allows a large space over the sand into which it can be allowed to rise when the mill is operating and a stabilised base is being pumped into the bottom of the mill for processing. This means of course, as for many other mills, that the pigment base will tend to occupy a space much greater than the voids of the milling media when at rest. The extent of space depends on the rate the premix is being introduced into the mill. The faster the pumping rate the higher the sand rises and the greater the ratio of pigment base to sand. The faster rate also means less friction with the sand in the mill because the greater amount of base means more lubrication with the sand. Therefore within the mill the temperature tends to drop with high throughput although the sand grating on the screen produces a certain amount of heat by friction. This, however, may be disregarded except for very heat sensitive materials. However, the main result of faster throughput is a reduced efficiency of milling. Therefore the rate must be adjusted primarily for milling efficiency and compromise only effected when absolutely necessary for any of the other parameters.

The vertical closed mills are in the main an extension of the open types but fully enclosed to enable the whole to operate under pressure. The screen enclosure still permits the mill media to move but this is more restricted in some models which have a slit type of separator acting as a dynamic separator. The main differences between the usual vertical closed and open mills is that sand is not recommended – glass or other synthetic types should be used instead – and the closed mills operate under pressure to enable bases of high consistency to be processed. Here again the initial charge may be left to the supplier but only as long as the initial trials are made because there are certain factors that should be established initially.

The usual recommendation for milling media for vertical closed sand mills of the larger type has been an equal mixture of 2 mm and 3 mm glass beads. It must be remembered that in these mills there is a large column of beads and consequently an extremely heavy weight at the lower end of the mill. This heavy weight must be rotated by the impellers for the mill to operate. The smaller the beads the greater the packing together and therefore the more effort required initially to start the mill, and one reason for the mixture of sizes noted above is that 3 mm beads were placed at the lower end. This allowed a less compacted condition than would be possible with 2 mm beads, and therefore an easier and freer start. However, the major point with freer starting is that all beads – especially glass – should always be left with ample lubrication, preferably with a resin solution that can be used in the next batch. It is dangerous to leave only a minute amount of lubrication because the heat evolved can volatilise the solvent and burn the resin, leaving a charred mass of resin and glass!

The horizontal mills have the great advantage that the weight of beads are not involved in the start of operations. The beads are in a horizontal cylinder, through the centre of which is the shaft actuating a number of disc or other types of impellers rotating in a vertical plane. Therefore gravity does not have to be overcome to anything like the extent when starting vertical mills.

The main characteristic distinguishing the horizontal sand (or bead) mill from the vertical models is the high milling media to premix ratio that may be designed. Some suppliers have indicated the range of suitable milling media apparent volumes to be between 70 and 90% of the total empty mill volume, the remainder of the space being occupied by the pigment paste passing through the mill. However, as the main advantage of the horizontal mill is the high ratio of milling media to pigment paste (the higher this ratio the better the milling action), the author considers that a mill charge of apparent volume equivalent to less than 75% of the mill volume would be unrealistic except for very special circumstances. Moreover a charge of more than 85% must be treated with caution because of possible problems with respect to pressure and temperature.

7.3.9.5 *Operating principles.* As the sand mill is a continuous mill it has not been found easy to design one in which stabilisation can occur during the processing of a base passing through the mill. One such mill was designed several years previously but most have remained such that they must be fed with a premix that has been stabilised.

Most premixes for sand mills are prepared on high speed mixers and this is satisfactory provided that the principles previously described have been followed. On the other hand, it must be stated that the ideal type of premix from a technical point of view would be prepared in a heavy duty mixer especially when organic pigments are being processed. Nevertheless, whichever method is used, the premix prepared with the proper attention to detail is stabilised to the consistency most suitable to the type of sand mill available. Ideally the stabilised premix will then be processed (milled or refined) through the sand mill in one pass. If more than one pass is necessary then, unless the pigments are particularly difficult to mill or there is some other known factor, the procedure adopted, including the formulation of the initial premixing, the stabilisation procedure and the sand mill charge (of mill media) should be reviewed.

With the horizontal mill there is a specific factor to consider and that is the pressure in the chamber as well as the temperature of the material as processed. The latter is determined (and therefore restricted) by the formulators and technicians. The former – the chamber pressure – must not exceed that maximum clearly defined by the manufacturers (usually 1.5 bar). Therefore provided there is no doubt that both the pressure and temperature are fairly low, when processing given materials, scaling up on a larger machine will allow production to be scaled-up in the same ratio. If, however, one is at or near the limits, scaling-up will be governed by those limits, viz. if a model is to be increased in size n times the production rate (as a result of pressure and/or temperature limits) will be increased by a factor $n^{2/3}$.

However, when considering the high-class automotive gloss enamels for which these mills are ideal it has been shown that the horizontal mills tend to perform two to three times better than vertical mills of the same shell size, provided the rheology is not abnormal.

It is understood that Draiswerke have models which are interchangeably vertical and horizontal mills because of their special design. These specific types must be regarded as being more like the horizontal mills described above than the vertical ones.

7.4 Formulating for optimum wetting and dispersion

7.4.1 *Introduction*

Pigments are usually received in the form of agglomerates of clusters of primary particles, which, in a liquid environment, could be separated by the application of a shearing action. The latter must of necessity be sufficient to overcome the cohesive forces within the multiparticulate structures. However, pigment manufacturing procedures may result in sintered aggregation. Provided that the latter is not too pronounced, a given amount of comminution is necessary before adequate separation can be effected. The ball mill is probably best suited to allow a given amount of comminution but it should be recorded that the ideal conditions for dispersion and comminution are almost mutually opposed. The high viscosity required for the shearing action for dispersion would tend to soften any hammering effect required for comminution. In practice the latter process calls for dry grinding or milling in water or other low viscosity material. Therefore, ideally, no comminution should be necessary because the pigments should have been satisfactorily treated at point of manufacture.

7.4.2 *Ball mills*

7.4.2.1 *The dispersion stage*. The composition of the pigment mill base charge should adhere as closely as possible to the guide lines in Table 7.2 or the relevant (ball mill) curve in Figure 7.2. The volume of this base should just fill the voids of the milling media while operating. Thus if the mill is loaded with a total apparent volume of milling media to 50% of the mill, the volume of the pigment mill base expressed as a percentage of the total mill volume would be

for spheres	20%
for ovoids	22.5%
for cylinders	17.5%

with a latitude not exceeding ± 2.5%. Reduced apparent volumes of milling media should be reflected in the reduction of pigment mill base, e.g. in mills with only 33⅓ % apparent volume of milling media the above figure would be reduced by a third.

Overloading the mill reduces milling efficiency and, although an apparently satisfactory result may be obtained upon lengthy milling, long term stability may be compromised.

Undercharging, although increasing milling efficiency, is also to be deprecated because increased wear and tear of the mill and milling media may result in unwelcome contamination.

Once the mill base formulation has been established and the requisite volume (suitable for the milling media) determined the dispersion exercise can be started. Liquids are loaded initially but a portion is retained to wash in the solid pigments. If a non-metal liner mill is used, a polar material if possible should be in all portions loaded into the open mill, to avoid or reduce static build-up. When all the additions are made the mill is clamped down and allowed to rotate. Of course the whole mill base may be premixed before charging.

The first couple of hours is purely a mixing action (unless the whole was premixed) to distribute the solids and liquids as evenly as possible to enable dispersion to occur. It is necessary, however, to determine whether cascading is occurring or whether by some mischance the consistency has been too great for satisfactory milling. This may be due to either a new material which has not been evaluated or a batch of material which has been prepared with a higher viscosity or oil absorption than usual.

On the basis that it is easier to reduce rather than increase the viscosity of a batch in a mill there is an allowance in the figures derived from the curve which would make the base have this bias. Generally there is no need to make any modification, but should slight thinning be desirable, solvent or solvent/resin solution should be used just to cause the change required. Many high oil absorption pigments produce undue and regular thickening with high yield points and thixotropy. The addition of resin at a number of stages is usually called for here. Whatever additions are made should be as little as possible, i.e. just enough to enable flow, even if this procedure has to be repeated a number of times; in this way a better dispersion is possible than if the total, finally added quantity, had been added initially.

This procedure is not very popular because of the labour involved in opening the ball mill, adding the resin, etc., closing the mill and restarting. For this reason, unless the most critical results are required, there is a tendency to add more liquid than necessary to make the base just flow, to avoid further additions.

If milling times exceed the 'norm', then, unless specific conditions have caused the problem and can readily be identified, the formulation and quantity of base must be checked together with the milling media size and quantity.

If first-class dispersion of organic pigments is required the results should be achieved within 48 hours. Only very exceptionally difficult pigments would require 72 hours. Apart from these cases, most materials should be readily milled within 24 hours or less.

7.4.2.2 *Stabilisation.* When the dispersion is completed stabilisation is necessary by adding resin, etc. to bring the formulation to that suggested in

Table 7.1. The number of resin additions and consequent rotation (to remill any flocculation) depends on the non-volatiles of the resin in the pigment base as shown in Table 7.4.

Table 7.4 Number of additions in let-down

Non-volatiles of resin in pigment base (%)	Number of additions of resin solution for stabilisation
40+	1
30+	2
25+	3 (with care, 2 possible)
20+	4 (with care, 3 possible)
15+	5 (with care, 4 possible)

Where a large number of additions is necessary, a small quantity of resin should be added at first and the quantity increased with each addition. It is necessary to rotate the mill after each addition for half to one hour in order to remill any flocculation. If at any one time the flocculation persists, then resin solution has not been added in sufficiently small quantities.

7.4.3 *Attritors*

7.4.3.1 *The dispersion stage.* The composition of the pigment milling base should be based on the suggestions given in Table 7.2 and on the attritor curve in Figure 7.2. This will ensure laminar flow for adequate shearing.

There are two methods of proceeding with the dispersion:

1. Some of the liquids are charged, the attritor is started and the solids added. The remainder of the liquid is used to wash in the solids while the machine is operating. This is a most suitable procedure provided that 'dusty' materials are avoided or are sufficiently carefully added to prevent contamination of other plant, or even parts of the attritor difficult to clean.
2. The whole milling formulation is premixed in a high shear rate mixer then added to the attritor either by pumping or via a chute. This method is most suitable for the larger type attritors and for repeated successive batches of the same composition.

7.4.3.2 *Stabilisation.* When the milling has satisfactorily reached the required dispersion, stabilisation is of course necessary. As the mill is an 'open' type the procedure is quite simple because the stabilising liquid can be added continuously, provided that the rate of addition does not exceed the rate of incorporation. This is of necessity slow at first but speeds up as the operation proceeds when the 'stable' base formulation for the requisite material is reached. The mill is then discharged and washed with the remaining liquids in the formulation, or another fraction of the same type is added for milling.

7.4.4 *Batch bead mills*

The composition of the pigment mill base must be prepared in accordance with Table 7.2 or the relevant curve in Figure 7.2 and the methods of charging would be similar to that suggested for attritors, that is:

- add some liquids, start operating, add solids and wash in with remaining liquids; or
- prepare the pigment mill base as a premix and add the whole to the beads in the mill.

As with other mills, the volume of mill base should be equal to the void volume while operating to obtain the most efficient dispersing action. Mill times vary with the raw materials, e.g. up to 15 minutes for readily dispersable materials, 30–40 minutes for most organic pigments and 60 minutes for blacks.

These times are exceeded only when abnormally difficult products are processed, e.g. transparent oxides. When the milling is satisfactory, stabilisation is accomplished, bringing the formulation to that of Table 7.1, in a manner similar to that described for attritors.

7.4.5 *High shear stress mixers*

Most of the liquids, based on the recommendations from Table 7.2 or the premixing curve in Figure 7.2, are charged and the machine started. Cooling may be necessary with some formulations as the pigment or other solid is added and incorporated. Initially the rate of addition is fast, but this is slowed to match the rate of incorporation. The power consumption must be noted to avoid cut-out and liquid added if necessary. Finally, if the procedure and formulation have been satisfactory the solids will have just been incorporated with the aid of the residual liquids while the power requirements would indicate the maximum available.

At this stage, as the machine continues to operate to disperse the particulate solid (or pigment), a difference in behaviour (and therefore a necessary difference in procedure) occurs depending on the material used as described below.

If a low oil absorption pigment is being dispersed there is unlikely to be any great change in viscosity which would vary mainly with temperature. If, however, the temperature had been increasing during the first stage of pigment addition, due to the increase in internal friction, it would have virtually reached a maximum by the time all had been added. Moreover cooling is usually proceeding. The only reduction in viscosity may be due to the retention in the interstices of pigment clusters of extra media which are then released. This is most unlikely to occur in the conditions appertaining but could be matched by unwetted internal surfaces absorbing this released material.

The dispersing action is therefore allowed to proceed for about 15 minutes, after which time no further perceptible improvement in dispersion would be likely to occur. Therefore stabilisation is effected by allowing the machine to continue to operate while the stabilising liquid is allowed to run into the mixer. This should be slow at first, ideally at a rate at which the liquid is being absorbed into the charge. This reduces the tendency to form an interface for any pronounced length of time, and as flocculation always occurs at the interface this procedure is an obvious further safety factor.

If the pigment used was an easily dispersing material, such as most of the titanium dioxide pigments available, it would be quite feasible that the pigment has been sufficiently well dispersed to satisfy even the most critical requirements (e.g. the automotive industry) and the stabilised pigment base would then be suitable for completion to the final paint without further milling. If, on the other hand, the pigment was not an easily dispersible type and required further processing, the stabilised base would be suitable for presenting to the continuous mill for further processing. In the latter case the continuous mill would yield the best possible dispersion only if the premixing had been accomplished in the specific manner prescribed.

If a high oil absorption pigment is being processed there is every likelihood of the viscosity increasing because such materials have small sized particles, which when subject to dispersion processes manifest large surface areas *in toto* which have to be wetted by absorption of the liquid. The removal of liquid for the wetting of the surfaces reduces the amount remaining for lubrication and flow. Eventually the consistency may become too great for the machine to operate, in which case further liquid must be added to maintain the operation. Here again only sufficient liquid just to maintain flow must be added (while avoiding motor cut-out or machine distortion) even if this addition is necessary at repeated intervals. This procedure is necessary to avoid reducing the viscosity to such an extent at any stage that very little shear stress is apparent.

Eventually, however, the continued addition is such that the concentration of pigment becomes too low and even the very high consistency maintained will not provide any satisfactory level of shear stress. At this stage, therefore, the material is brought to a stable type of formulation and presented to another mill for more efficient milling, unless the dispersion is satisfactory. The mill to which the premixed paste is transferred is either one of the continuous sand mill types or the roll mill family.

At this stage it would be helpful to discuss the main reasons why a higher non-volatile vehicle is suggested for the premixing process in a heavy duty mixer (i.e. low shear rate/high shear stress machine) than is recommended for a high speed mixer (to be discussed next).

In point of fact the dispersing action during premixing is equally successfully accomplished in both types of mixers. However, it is in the stabilisation of the premix that trouble may occur.

Firstly it has been emphasised that flocculation readily arises at the interface between dissimilar materials (e.g. a pigmented vehicle and the same vehicle not pigmented). The longer the interface lasts the more time there is for flocculation to occur and last! The slow deliberate mixing of the heavy duty machine tends to create the conditions to favour this unless added at a very slow rate initially as stated.

Secondly, however, the addition of a resin to a pigmented resin with similar non-volatiles will not reduce the overall consistency as markedly as adding the resin to a pigmented low solids resin. In the latter case the consistency will soon reduce to a stage where shearing action is poor, so any flocculation formed has little chance of reincorporation by dispersion action. In the first case of resin addition, however, the reduction in consistency is not so marked and would therefore be sufficient to permit the redispersion readily.

7.4.6 *Low shear stress (high shear rate) mixers*

The premixing or mill base formulation is determined by referring to Table 7.2 or the curve for premixing in Figure 7.2. The liquids are charged into the tank of the mixer and should cover the disc reasonably well. The mixer is started. If of variable speed, a low speed should be used initially. The pigment or other particulate solid is charged while operating. Increasing speed will be necessary to incorporate the pigment at a satisfactory rate. Incorporation could be helped if the discs can be raised and lowered while operating and also if some of the liquids had been retained to wash in the last of the solids, since the last portions are always more sluggish to incorporate.

Once the stage has been reached when all the particulate solid has been incorporated it is often advisable to scrape down the sides of the tank to incorporate any powder that has settled there. An alternative procedure is to wash the sides down with the minimum quantity of solvent possible, which should have been omitted from the initial charging for this purpose.

The mill will then be allowed to run for about 15–20 minutes, after which time, provided that the charging has been satisfactory, no further real change is likely to occur with respect to the dispersion of the particular solid. Care must be taken that the mixer is operating satisfactory, i.e. a 'doughnut roll' is evident in the mix with adequate movement round the sides. If these features are absent the formulation being presented to the mill is inadequate and should be modified either to increase or decrease the consistency.

As with the heavy duty mixers discussed previously, viscosity may increase when processing organic pigments or other high oil absorption solids. Again the minimum quantity of liquids should be added to maintain flow. Even if frequent additions are necessary this is better than a larger single addition for dispersion efficiency.

It will be noted that the mix rapidly becomes heated because of the internal friction. Theoretically the heat should reduce viscosity. In actual fact this is so, but with the high oil absorption pigments the small particle sizes present large surface areas to be wetted and therefore cause a reduction in the resin or other medium available to enable flow. Moreover, the rate of wetting is increased at elevated temperatures. With low oil absorption pigments, however, the particle sizes are much larger and therefore the surface area to be wetted is not so great. Here the elevated temperature would tend to reduce the consistency.

This is made use of in simple formulations including an easily dispersible titanium dioxide/resin solution, an additive such as soya lecithin and white spirit. The titanium dioxide is dispersed in the white spirit and soya lecithin only (hence zero resin non-volatiles quoted in the table) to the extent that the mix may contain up to 80% titanium dioxide! This would normally be a highly dilatant mix, but because the temperature rises in the operation to nearly 70°C the resin on expansion causes a vast reduction in consistency and becomes virtually Newtonian in character. Of course the viscosity is still quite high, enabling the dispersion process to occur. It is imperative, however, that the mixer should not be stopped before full stabilisation is effected, otherwise cooling will cause the dilatancy to be re-established and it will be impossible to restart the mixer.

Stabilisation should be effected in the usual manner, i.e. by adding the requisite resin solution while the mixer is operating. The addition should be slow at first, to reduce any effect of interfacial reaction, but as the consistency is reduced, the rate of addition may be increased.

It may be pointed out that the liquid non-volatiles for the high speed premixing of low oil absorption pigments are quite low (and even zero, as pointed out above). This is because the low viscosity liquids (solvents and highly solvated resins) are better wetters than the more viscous high solids materials and usually contribute to producing a well-wetted, well-dispersed product. However, at one stage the introduction of new grades of titanium dioxide specifically designed for ease of wetting produced rather unexpected problems. The excellent wetting characteristics caused the titanium dioxide to absorb the low viscosity liquid with such rapidity that pellets were formed enclosing quantities of titanium dioxide that were not available for dispersion because the pellets themselves would not disperse in a high speed mixer and often blocked the entry screen or valve of continuous sand mills.

The procedure adopted to overcome this problem was to ensure the dispersion of the titanium dioxide before it was well wetted. This entailed the use of a high non-volatile liquid which did not wet well but tended to shear better, instead of the low non-volatile liquid originally recommended. This indicates the necessity to maintain flexibility in the application of the well-tried dispersion techniques.

7.4.7 *Other high speed mixers*

The best formulations for these depend on the type of machine.

Rotor stator mixers should have a formulation based on high speed mixers but of a slightly lower viscosity to allow for the complicated motion of the base being processed through the holes or slits in the stator. Apart from this the procedure should be identical.

Other combinations of mixers should be treated as far as possible as for high speed mixers. Any modification depends on the construction of the plant and may require some increase in the quantity of liquid non-volatiles for lubrication of motion. However, the main feature must be to maintain as high a viscosity as possible until the dispersion is satisfactory. At this stage stabilisation follows at a rate not exceeding the rate of incorporation of liquids. This is slow at first but becomes faster as the operation proceeds.

7.4.8 *Continuous mills*

Attempts to stabilise a base being processed in a continuous mill have not so far been very satisfactory. Therefore in order to conclude with a satisfactorily milled stable base from continuous type mills, it is advisable to feed these machines with a base formulated for stability as in Table 7.1 and Figure 7.2. These latter indicate the maximum pigmentation, therefore it will be necessary to formulate the remaining part of the base to obtain the highest viscosity possible suitable for the machine. Generally speaking the higher the non-volatiles the higher the viscosity. However, a very important aspect of the whole process is to ensure that premixing for continuous mills should be of the highest efficiency, because the more the effort applied in the premixing stage, the easier and better is the dispersion in the final dispersing stage. A poor premix will require repeated passes through the continuous mill and even then may not achieve the finally required result.

The ideal premixing machine is a high shear stress type of mixer of heavy duty construction to enable a very high viscosity base to be processed. The result would be most suitable for processing on the roll mill. However, reduction in viscosity to a suitable consistency would be necessary for sand mills. It is cheaper to use high shear rate mixers for sand mill premixes and very good results are obtainable provided due attention to high viscosity as far as possible is observed. Similarly high shear rate mixers could be used for premixing for triple roll mills if high shear stress mixers are not available. In all cases of premixing the process should be as efficient as possible to provide the continuous mill with the most suitable base for final milling.

7.5 Choice of equipment

7.5.1 *Introduction*

The basic criteria for manufacture are:

- To achieve the quality required.
- To ensure repeatability of quality.
- To ensure the profitability.

All dispersing machines will be able to ensure repeatability of quality, provided the principles previously discussed are followed. Most important is the necessity to start all dispersion exercises as though the best possible result was required. The process can then be cut off at the stage the necessary quality is reached. It is a mistake to be lax on these matters just because a dispersion of a quality lower than the highest quality is desired.

However, not all dispersing machines now in operation will obtain the highest quality. As an example, the high shear rate mixers will not generally cope with pigments unless they have been specifically treated to enable easy dispersion. The reason is that the machines are low shear stress types which are not suitable for the high viscosity required for adequate shearing action. High viscosity pastes are indeed processed on these mixers but only when the paste is hot (as a result of internal friction during mixing) which effectively reduces the consistency.

In these cases reduction to low viscosity before the process is complete is essential. It is necessary for the process to be properly costed rather than rely on a blanket type of costing. The various details required for a typical range of products must include depreciation, power consumption, and man-power costing with respect to preparation, charging, milling, testing, reduction, discharge and washing. The exercise must include the profound difference in costs when different sized mills are considered. This is more pronounced in batch mills than in continuous mills. An effective costing exercise for a range of mills in one particular factory gave a result in the following order (cheapest first).

1. High shear rate mixers.
2. Large horizontal continuous bead mills.
3. Large steel ball mills.
4. High shear stress mixers.
5. Batch bead mills.
6. Pebble ball mills.
7. Roll mills.

This order could possibly be somewhat different for various factories depending upon specific practices. However, the ranking determined by each factory could enable the most economical procedures to be devised. For example, the difference in cost could be established for the choice of a cheap coarse extender which will necessitate the use of a ball mill for a degree of comminution or a more expensive and finer extender which could readily be processed in a high shear rate mixer. It must be emphasised that each factory must determine the order for its own procedures and redetermine where variations in procedures occur.

7.5.2 *Emulsion paints*

All emulsion paints should ideally be processed on a high shear rate mixer. It would be false economy to use cheap coarse extenders which have to be dispersed in a more expensive machine such as a pebble ball mill.

7.5.3 *Alkyd gloss*

If these are tinted titanium dioxide paints then with the proper choice of easy to disperse pigments the high shear rate mixer is again the most suitable one to employ. The pigment is often best dispersed in a mixture of solvent and the requisite surfactant. Occasionally, however, a minimum of resin is required in specific formulations. The processing should be completed within 10–15 minutes mixing time and reduction is effected by adding the resin remaining with other materials with continued mixing. The additions should be slow at first with the rate increasing to match the rate of addition.

If the pigmentation includes a quantity of other and less easily dispersible pigments the procedure should be the same but, after reduction to a stable base type formulation, the whole should be passed through a continuous mill of the sand bead mill type. The rate of passage through this latter mill will depend on the ease of dispersion of the additional pigments.

7.5.4 *Stoving enamels*

Very high gloss black enamels, such as for automobiles, would be most readily milled satisfactorily and economically in large steel ball mills. The fairly low pigmentation would enable large batch sizes to be prepared in 48–72 hours dispersion time.

Highly coloured types are best prepared in continuous bead mills but specifically small batches would be most suitably produced on batch bead mills or attritors.

7.5.5 *High viscosity types*

These are used in silk screening and may be classified more correctly as inks, and would require premixing in high shear stress mixers with subsequent refining over a roll mill.

7.5.6 *Speciality types*

The role of the pebble mill has not completely vanished. Many companies prefer them because their operators are far happier with old simple plant than with modern types. There are still some paints, however, which require such mills, notably

(a) Paints which have to be absolutely free from any contamination from the colour point of view
(b) Formulations which require a given amount of pounding for certain texturing finishes
(c) Materials which require processing in an ambient absolutely free from metallic contamination
(d) Pebble ball mills have also been used to process materials which should be kept away from air or oxygen as far as possible.

Bibliography

O. Absolon, Rationalisation of dispersion of organic pigments, *Plaste Kautschuk* **26** (12) (1979) 708–9.

Anon., Co-ball mill, *Pitture Vernici* **64**, (12), (1988) 58; 53–7 (Italian).

K. Apostolakis, Mechanical equipment for the industrial production of paints, *Proc. 1st Symposium on Paints* (Athens 1987), 173–89 (Greek).

P. Banerjee, Criteria for the selection of dispersion equipment, *Paintindia* **38**, (12), (1988) 29–32.

I. Berg, Dispersion techniques in the paint and allied industries, *Polym. Paint Col. J.* (1983, Suppl. to Feb 23rd issue).

F. Bilgieri, Optimal efficiency of titanium dioxide millbase formulation, *Proc. XVth FATIPEC Congress*, (Amsterdam 1980) **1** 449–60 (French).

A.J. Boulton., Latest developments in horizontal grinding mills, *JOCCA* **69**, (4), (1986) 101–2.

B. Burgess, Dispersion history: from batch bead mill to edge runner, *Coatings Mag.* **7**, (1) (1985) 106.

B. Burgess, Disperser, mill review:choosing shaft arrangements. II. *Coatings Mag.* **7** (No. 2) (1985) 58–60.

B. Burgess, Coatings mixing equipment. I. Modern dispersers handle chores formerly left to mills, *Coatings Mag.* **7** (4), (1986) 50.

B. Burgess, More attention should be paid to let down mixer equipment, *Coatings Mag.* **7**, (6), (1986) 24.

W. Carr, Dispersion: the neglected parameter, *JOCCA* **65**, (10) (1982) 373–82.

W. Carr, D. Patterson and M. Tonge, Effect of pigmentation level on efficiency of triple roll milling, *JOCCA* **63** (11) (1980) 438–45.

Chemineer Ltd., Static mixers, *Polym. Paint Col. J.* **179**, (4229) (1989) 28–9.

B.C. Clarke, Mixing: a review of basic principles and developments, *Polym. Paint Col. J.* **175**, (4152) (1985) 699–70.

L. Columbo, Revolutionary Italian technology (for pigment dispersion), *Pitture Vernici* **65**, (7) (1989) 27–9.

A.C.D Cowley and M.R. Gallon, Who needs wetting agents?, *JOCCA* **71**, (10), (1988) 310.

L. Cutrone, High solids millbase design for high speed dispersion, *J. Coatings Tech.* **56**, (716) (1984) 105.

F.K. Daniel, Getting the most out of a high speed disperser, *Am. Paint J.* **66** (No. 46) (1982) 60–5.

F.K. Daniel, The obstacle course from millbase to finished coating, *J. Coatings Tech.* **56**, (708) (1984) 38.

K. Engels, Bead mills with annular cross section grinding chamber, *Farbe Lack*, **91**, (12) (1985) 1174–9.

K. Engels, Pearl mills in 1986: an overview of the pearl mill market; *Farbe Lack* **92**, (8) (1986) 727–33.

K. Engels, Microbeads in beadmills, *Farbe Lack* **94**, (5) (1988) 359–64.

R. English, Recent progress in milling processes, *Proc. 14th AFTPV Congress*, (Aix-les-Bains 1982) pp. 259–62.

L. Farsky, Ultrafine dispersing in horizontal ball mills, *Chem. Prum*, **36** (8) (1986) 404–7 (Czech).

M.L. Feldman, Controllable variables during dispersion manufacture, *Am. Paint J.* **72** (26), (1987) 38–44.

C.A. Giudice, J.C. Benitez, V.J.D. Rascio, and M.A. Presta, Preparation of paints in ballmills. Influence of variables associated with change of scale, *CIDEPINT Anales* (1980) pp. 33–60.
T. Gondoh, M. Toh and Y. Murakami, Power consumption and discharging characteristics of batch horizontal sand mills, *J Jap. Soc. Col. Mat.* **61** (11) (1988) 593–8.
T. Gondoh, M. Toh and Y. Murakami, Behaviour of grinding media in sand mills, *J. Jap. Soc. Col. Mat.* **62**, (4) (1989) 202–7.
J.A.Graystone, Optimising pigmented intermediates, *Proc. Paint R.A. 8th Internat. Conf.*, 'Optimising Paint Formulation' (1988).
G. Gruhn, L. Pollmer and B. Henze, Optimisation of dispersion processes in paint production, *Plaste Kautschuk* **31** (1) (1984) 32–5 (German).
N. Harny, M.F. Edwards and A.W. Nienow, *Mixing in the Process Industries*, Butterworths Series in Chemical Engineering (1985) 375 pp.
A. Heuss, Microsphere bead mill, *Pitture Vernici* **61** (8) (1985) 46–8 (Italian).
J.E. Jarvis, Mixing of solids, powders and fluids, *Chem. Engr.* **(358)** (1980) 477–81.
J. Johnson, A. Szegvari and M. Li, Attritor grinding and dispersing equipment, *Polym. Paint Col. J.* **(172)**, (4075) (1982) 459.
M. Jonatpour, The evolution of the paint mill – Industrial Finishing **64** (1), (1988) 24–5: *NPCA Abs Rev.* (567), (1988) 25.
S. Komori and Y. Murakami, Effect of highly viscous recirculating flow on particle concentration in a two-roll mill, *Powder Tech.* **57**, (2), (1989) 119.
R.M. Kritzer, Mechanics of dispersion and deaeration, *Paper Age* **104**, (4) (1988) 19.
R. Lapasin, D. Mallamiaci and E. Medaglia, Rheological properties of millbases and energy consumption in dispersing operation, *Proc. XVth FATIPEC Congress* (Amsterdam 1980) **1** 433–48.
J.C. Leveque and A.Ozaneaux, Use of statistical methods to control the dispersion of pigments, *Double Liaison* **29**, (316/7) (1982) 18–24 (French).
G.R. Liebert, Advances in the milling of high solids coatings, *Mod. Paint Coatings* **75** (5) (1985) 58.
D. Mallamaci and E. Medaglia, Rheology and energy aspects of dispersion in microelement mills, *Ind. Vernice* **34** (10) (1980) 3–18.
B. Mayo, J.P. Pfau and R.E. Sharpe, Monitoring pigment dispersion using the free volume microprobe, *J Coatings Tech.* **59**, (750) (1987) 23–7.
J. Medek, J. Klohna and K. Hruban, Some possibilities for innovation in stirring equipment, *Chem. Prum*, **35**, (2) (1985) 61.
W. Muller, Use of high speed mixers provided with dispersion discs in the paint industry, *Plaste Kautschuk* **33**, (1) (1986) 32–4 (German).
Y. Murakami *et al.,* Dispersion and classification effects in a three roll mill, *J Coatings Tech.* **57**, (727) (1985) 73–81.
Y. Murakami *et al.*, Classification and dispersion performances of a three-roll mill, *J Jap. Soc. Col. Mat.* **59**, (7) (1986) 385–90.
J. Oyarzun, Dispersibility of pigments as criterion for determining optimal mill base formulation, *Proc. 16th FATIPEC Congress* (Liège 1982) **IV**, 93–116 (German).
Paintmakers Association of Great Britain, *Guidelines on the Safe Use of Ball Mills, Mixers and Dispersers*, Paintmakers Association, London (1981) 10 pp.
G.D. Parfitt, Dispersion of powders in liquids, *Kirk Othmer Encyclopeadia of Chemical Technology*, 3rd edition, Wiley Interscience (1984) Suppl. pp. 339–71.
I.C. Quackenbush, Points on picking the right media for small media mills, *Am. Paint J.* **67**, (27) (1982) 46.
J.M. Rackham, How the titanium dioxide industry can help the coatings industry save energy, *Proc. XVth FATIPEC Congress* (Amsterdam 1980) **1** 279–98.
C. Reitter, Using numbers, not guesses, in making paint, *Am. Paint J.* **69**, Part I (44) 48; Part II, (45) (1985) 56.
A.G. Sakar, Equipment for dispersing pigments in binders, *Lakokras. Mat.* (3) (1983) 50 (Russian).
L.F. Sanguinetti, Paint manufacture, *ACS Symposium Series* 285, (1985) 1297–314.
G. Schall, Developments in the field of wet grinding and dispersing, *Am. Inkmaker* **63**, (2) (1985) 52.
M.K. Schieritz, Horizontal ballmills, *Farg Lack* **29**, (6/7) (1983) 153–7 (Danish).
B.K. Sharma and N.A.R. Falla, Evaluation of new methods of paint milling, *Paint R A Tech. Rept.* TR/3/85, 44 pp.

L.A. Simpson, Effect of pigment dispersion on opacity. *Proc. XII Internat. Conf. in Organic Coatings Sci & Tech.* (Athens 1986), 343–58.
J. Sir, Z. Lecjaks, J. Machac and A. Zakova, Static mixers, *Chem. Prum.* **35** (3) 113–17 (1985) (Czech).
C.C. Tatman, Solvent bourne millbase development for high speed dispersion of titanium dioxide, *J Coatings Tech.* **53**, (681) (1981) 57–65.
J.E. Teirfolk, Flow picture in the mixer during dispersing of coating pigments, Paper *BPBIF Conf. on Coating for the 80's* (Slough 1980); *Abs Bull. Inst. Paper Chem.* **52**, (81) (1982) Abs. 8671.
F. Vock, K. Warnke and K. Sollik, Improvements in the dispersion process using a new mill with a stirrer, *Farbe Lack.* **93**: Part I (11), 905–10; Part II, (12), 995–1000.
M. Ward, Rheology of mill bases for optimum dispersion of titanium dioxides, *Australian OCCA Proc. News* **19**, (4) (1982) 20–3.
B. Westerlin, High speed mixers – 'a theme with variations', *Farg Lack* **29**, (6/7) (1983) 149–52 (Swedish).
J. Winkler and Dulogh, Dispersing pigments in paints with a continuously operating attritor, *Farbe Lack* **90**, (4) (1984) 244–50 (German).
J. Winkler, E. Klinke, M.N. Sathyanarayana and L. Dulog; Theory for the deagglomeration of pigment clusters in dispersion machinery by mechanical forces. I–III, *J. Coatings Tech.* **59**, (754), (1987) 35–41; 43–53; 53–60.
Y. Yamawaki, Introduction to pearl mills, *J. Jap. Soc. Col. Mat.* **55**, (4) (1982) 238–44.
C. Zoga, Dispersion and milling methods to increase plant productivity, *Mod. Paint Coatings*, **79**, (5) (1989) 48.

8 Selection of test methods

RICHARD WOODBRIDGE

8.1 Introduction

Having formulated a single or series of experimental paints, they must then be evaluated to establish whether the expected results have been achieved. This involves selecting appropriate test methods and carrying out those tests in as controlled a manner as possible. This may appear to be quite straightforward but, unless positive thought is given to both aspects, it is quite possible that considerable effort will be wasted by using inappropriate methods, substrates or conditions, making the results, at best, questionable and, at worst, valueless. As in all other aspects of the formulation process, the technologist is encouraged to think critically of what is to be achieved before starting work.

Ideally, test paints should be applied, dried or cured and tested under conditions which reproduce as closely as possible the conditions likely to apply in practice. This statement may be seen as being in conflict with the need, for the purposes of reproducibility, for standardising the method, wherever practicable. In fact, a degree of compromise is required, in which the technologist attempts to eliminate sources of variability whilst trying to achieve the maximum degree of realism within a method capable of being carried out in the laboratory at acceptable cost. It may be thought that such an approach is applicable more to 'subjective' tests, but consideration of the basic principles being advocated can also identify their applicability in some aspects of 'objective' methods. To illustrate the principles involved some examples are considered; first ways in which sources of variability may be eliminated and then examples of moving towards greater realism.

8.1.1 *Sources of variability*

One of the first variables likely to be encountered within a series of paints to be tested 'in parallel' is their history. Variations in method or scale of manufacture, or in length or conditions of storage before testing are all likely to influence results of tests – some being affected quite significantly by such differences. An obvious example is in the field of viscosity measurement, particularly at low shear rates, where the result will be influenced by the amount of structure which has built up during storage. Such variations can be reduced significantly by breaking down the variable levels of structure, using standardised conditions of high shear mixing, and leaving all samples to stand for a predetermined period before making any measurements. Such precautions would, of course, not be necessary if measuring high shear viscosity with a cone and plate viscometer, providing the viscosity reading is allowed to stabilise. Using the same example of viscosity measurement, the degree of variability in results can also be reduced by selection of the most appropriate test method. How often have there been attempts to measure the viscosity of a thixotropic product by stirring it through an efflux type viscosity cup? Use of a Krebs Stormer Viscometer would be much more appropriate when only a single point measurement is required, say, for control purposes, whilst ideally a variable speed rheometer should be used to obtain a rheogram over an appropriate range of shear rate or shear stress. Other properties which may be influenced by the short term history of the product are gel strength, gloss or sheen level, level of entrapped foam, coalescence, drying of alkyd-based paints, sagging tendency, etc. When time allows, it can be advantageous to let experimental samples stabilise for 24–48 hours before carrying out extensive testing.

A further example of potential variability is in the application of test films of equal thickness or at their natural spreading rates. Various techniques are used to try to establish constant film thicknesses, but they are full of pitfalls. Use of doctor blades with preset gaps will give variable results depending on the rheology of the paint or on the speed of drawdown. For a given gap setting, a wedge-shaped applicator will lay down a much thicker film than an equal gap 'square section' blade. Increased shear during application, simulating more closely conditions during application by roller or brush, can be obtained by using a wire-wound bar coater, but for reproducible results it is essential to maintain absolute cleanliness of such wire-wound bars. A common practice of dipping them in appropriate solvent and wiping dry after use very quickly leads to a build-up of pigment between adjacent wires, reducing the amount of paint spread. Experience suggests the regular discipline of soaking in strong solvent and brushing through the grooves with very fine brass wire brushes. A ×10 magnifier will quickly establish whether such coaters have been maintained in usable condition. With any such coater or doctor blade, reproducibility is greatly enhanced by the use of a vacuum plate to hold the test card and a constant speed applicator with standard loading on the bar.

Variation in film thickness, when using a 'BK' cube applicator for application of films for drying tests using a BK drying recorder, was found to be one of the factors responsible for variability in results during an investigation into the mechanism of drying, co-ordinated by the Paint Research Association at Teddington, England. However, a more fundamental influence on results was the nature and state of the needle. Extensive collaborative testing by industrial sponsors of the project, in which the cleanliness and position of the needle were carefully controlled (a practice unfortunately not followed by many users), also failed to give results within acceptable standard deviations.

Further studies on alternative needles showed that a PTFE-coated ball-tipped needle gave significantly better reproducibility than the standard Beck Koller needle. An alternative technique which was shown to be capable of providing reproducible data was developed and is discussed under the heading of 'drying tests'.

8.1.2 *Towards greater realism*

For convenience, many tests are carried out on steel, tin or glass panels or on card, which may be sealed or unsealed, as in hiding power contrast charts. Whilst such established substrates may be fully acceptable in many cases, they may be quite untypical of the substrate to which the paint is normally applied. In attempting to move closer to a typical substrate the technologist is faced with the problem that many substrates, in practice, are extremely variable, quite apart from their unsuitability as test substrates within the laboratory. However, it is important to recognise the variability in surface which may exist in practice since it may significantly affect the paint's performance or appearance. As an example, the colour uniformity of a flat latex paint may be influenced by variations in the absorption of the substrate. Plaster applied *in situ* will vary considerably in its absorption, depending on grade, the degree of trowelling carried out in finishing, speed of drying, additions of lime, the underlying substrate and other factors. Equally, wide variations in absorption may exist between different grades of factory-produced building boards, or joinery grade or cladding timber. Pretreatment or precoating of such materials will also influence the nature of the surface to be painted. Coatings designed for use on specific surfaces should logically be tested, wherever possible, on those surfaces. In the case of general purpose coatings, it is unrealistic to test on every potential substrate, but the range of conditions likely to be encountered should be recognised and it should be ensured that the normal extremes of conditions are catered for within the tests carried out. Apart from colour uniformity, coalescence of latex paint films, sheen, adhesion and porosity are just some of the properties which will be influenced by the speed at which the film dries, due to variations in absorption of the substrate. When such variations are likely to be encountered within adjacent localised areas of a substrate a standardised test surface should be established which reproduces those extremes by using a partially sealed absorbent test panel.

Drying conditions can also vary widely in practical paint application, in temperature, humidity and amount of air movement over the substrate. Paints likely to be applied under such varying conditions should be tested under the range of conditions likely to be experienced in practice. This may normally be recognised in carrying out drying tests, but is less frequently recognised when testing for other properties which can equally be affected by such variations in conditions. As an example, when preparing panels for exposure testing, it is normal practice to prepare them under ideal laboratory conditions and often to leave them to age for 1 week before placing on the exposure racks. Yet, in practice, the potential life of the coating can be influenced very much by the conditions existing during application and drying. High humidity, causing excessive delay in the drying of latex coatings, can adversely affect coalescence and hence film integrity, and significantly reduce their durability compared with films of the same product dried under 'normal' laboratory conditions.

Mudcracking is another example of a paint defect which can occur when a paint such as an exterior masonry coating is applied to a rough or textured surface, or a flat latex coating to a textured lining paper. Such uses often apply, yet the normal laboratory testing of such finishes is on smooth panels. Use of a 'standard' surface which gives typical 'pockets' of paint or controlled variation in film thickness assists in formulation of products free from such defects.

Most experienced technologists will recognise situations where a paint defect or problem has forced them to identify or develop a new test method to help in overcoming the problem. The aim, however, should be to anticipate such problems and ensure a trouble-free product before going into production. The more the ultimate conditions are recognised during the development stage, the more likely it is that the ideal will be achieved in practice.

8.1.3 *Choice of method*

In many cases, the tests to be carried out and the method to be used are specified by the user or within industry/national specifications. Wide variations exist within such published methods and it may be many years before harmonisation is achieved. Until such time, it is natural to use the methods which are accepted in a specific market or which have been established over the course of time within a company, providing critical review confirms that they are appropriate and realistic in today's conditions. It would be unrealistic to attempt to promote one particular method, in detail, against others already established. The main objective within this chapter is to indicate which tests should be considered within any particular investigation and to indicate in broad terms the approach which may be taken.

8.2 Properties in the can

One of the first judgements of the quality of a product is its appearance in the can when first opened. This will be influenced by its viscosity characteristics, storage stability, the choice of a suitable container and, in the case of in-store tinted products, its colourant acceptance. The relative importance of these different properties will depend on the nature of the product and its intended market, which will determine which tests are significant and which may be omitted. For example, an industrial product produced specifically for one customer may have a sufficiently high turnover to make storage testing irrelevant, whilst in the case of a tinting base for retail distribution, all tests should be carried out to ensure that the product reaches the end customer in usable condition and retains reasonable properties during the time on his shelf, before re-use for 'touch up' purposes.

8.2.1 *Viscosity characteristics*

Many instruments and techniques exist for the measurement of paint viscosity, some suitable only for 'one point' measurements, whilst others give a detailed viscosity profile over a range of shear rates or shear stress. Most have their uses, providing that their limitations are recognised and that a suitable instrument is selected for any specific investigations.

For low to medium viscosity paints with Newtonian characteristics, such as many products used in the industrial paint market, the long established flow cup is quite adequate. Providing adequate attention is paid to selection of the right cup for the viscosity range being tested, use in a stand with the axis of the orifice in a vertical position, and regular calibration to establish that the orifice is neither partially blocked with dried paint nor worn to a larger diameter by over-enthusiastic cleaning, this technique is capable of giving satisfactory reproducible readings for development or control purposes.

For products showing a degree of thixotropy or structure, more sophisticated techniques must be used. For control purposes, one or more 'single point' measurements are acceptable, such as use of a Krebs Stormer viscometer, ICI rotothinner or Brookfield viscometer at lower shear rates and use of a high shear viscometer such as the ICI cone and plate model to give a measure of the brushing/roller viscosity. As the degree of structure or gel increases, as in the so-called 'solid' paints, it becomes more important to consider a full rheological plot using an instrument such as the Feranti Shirley, Contraves or Carrimed viscometers. Even with such instruments it may be necessary to carry out more than one test, using different conditions, to give a more detailed picture at both the low and the high ends of the viscosity profile. Generation of a hysteresis loop at very low shear rates may be a necessary discipline, even for control purposes, on products where the strength of the gel is critical in affecting such properties as roller loading. Temperature control is extremely important in any viscosity measurement, particularly when trying to relate measurements made with more than one cup, bob or cone in a rotational viscometer, where temperature limits of ± 0.1°C are necessary.

8.2.2 *Storage stability*

The first aspect of this to consider is *viscosity stability*. Repeat measurements after a period or periods of time are made using the same conditions and instrument as when first measured. It is common practice to re-measure viscosity after 1 week, 1 month and after 3 and 6 months storage at both room temperature and at an elevated temperature (30°, 40°, 43° and 50°C have all been quoted). Use of elevated temperatures to accelerate storage testing should be used with caution, since it can produce effects quite untypical of room temperature storage. This will depend on the formulation, and, providing correlation has been established by past experience, storage at elevated temperature is a valid method of anticipating storage stability.

Synaeresis and *pigment settlement* are other defects which may be accelerated by storing at elevated temperatures or by temperature cycling, i.e. by storage overnight at the higher temperature and at room temperature during the day. Low speed centrifuge has also been used as a good prediction technique but again it is wise to correlate results on specific type paints before regarding the technique as infallible.

Freeze/thaw stability uses a variation on the temperature cycling test, using a refrigerator at −10°C or below for overnight storage. Paints are inspected daily before replacing in the refrigerator to establish how many cycles can be tolerated before excessive granulation or gelation takes place. In making such daily inspections the sample should be disturbed as little as possible to avoid influencing the result.

Resistance to attack by microorganisms is also tested by the incubation technique, assessing at regular intervals for loss of viscosity or development of putrid odours.

Loss of dry of alkyd-based paints on storage, or development of *gassing* in paints containing aluminium pigments can also be assessed on the incubated sample. In the case of loss of drying characteristics, results are likely to be more detectable if the comparison of drying is made at low temperature. When assessing the retention of *colourant acceptance* the results may be more significant if the incubated sample is tinted while still warm.

pH stability of latex paints is also assessed after progressive periods of incubation at elevated temperature.

8.2.3 *Degree of dispersion*

The most common method for establishing degree of dispersion or fineness of grind is the Hegman Gauge. While this is quick and simple to use, it really only gives a measure of the presence of oversize particles or 'bits' and is not satisfactory for assessing whether the full potential has been realised from the quantity of titanium dioxide used in the formulation.

More recently, an extremely useful technique has been developed by Tioxide UK Ltd to measure the flocculation gradient of a paint. The specially designed 'Flocculation Gradient Monitor' uses a monochromatic source of infrared radiation to irradiate a dry film of the paint under test, positioned within an integrating sphere. The percentage of back scatter is measured and related to the dry film thickness. For film thicknesses up to 30 μm this relationship is constant, giving a straight-line graph when plotting percentage back scatter against film thickness. The gradient of the line is referred to as the flocculation gradient and is a quantitative measure of the degree of dispersion or flocculation. The lower the value, the better the degree of dispersion. A simple cell has also been devised to make the assessment on wet paint, which can give the possibility of assessing the degree of flocculation which takes place during the drying process.

8.3 Application characteristics

Whilst this may be one of the more subjective assessments made in the evaluation of paints, a number of techniques are available to make more objective measurements to eliminate the problems of personal judgement and reproducibility.

Such techniques, however, cannot fully replace the practical assessment of performance which, for the sake of reliability, should be on as large a scale as possible. Ultimately the most reliable assessment is the full scale practical trial which inevitably brings into play all the various factors such as brush/roller loading, spreading rate, lapping, cutting in, etc., which together influence the overall ease of application. Equally in the case of application of industrial finishes, the final proof of acceptability is the trial on the plant, where ambient conditions such as temperature, concentration of solvent vapour, etc. may be very different from conditions in the laboratory on a small scale.

8.3.1 *Decorative finishes*

With such finishes the main concern is establishing the ease with which the product can be spread uniformly to the required film thickness and free from surface defects over either large wall surfaces or smaller, possibly intricate, mouldings. Whilst brush or roller are the traditional tools, pads may now be considered by the DIY user, whilst the professional decorator may also use conventional or airless spray. The main property influencing ease and speed of spreading is the paint's high shear viscosity (at 10 000 s^{-1}). The optimum viscosity for a particular paint will depend mainly on the size of the application tool; for wall finishes applied by roller or larger wall brush a high shear viscosity of 80–120 mPa s (centipoise) is normal. The higher the viscosity within this range, the lower will be the natural spreading rate with proportionate increase in film thickness. For finishes designed for use on trim, such as windows, doors, etc., smaller brushes or pads are used, calling for an increase in high shear viscosity to 250–350 mPa s (centipoise) .

Spreading rate is thus influenced by the combination of the paint's high shear viscosity, the actual tool used (including size, construction, amount of wear and condition) and the skill and strength of the decorator. Average or normal spreading rates are often quoted, and this is best established in practical tests on as large a surface as possible using typical applicators. It is common practice to reduce the results obtained from such controlled tests by approximately 20% to achieve a recommended rate. Such large-scale application tests are also the best guide to *lapping characteristics*. Whilst lapping is influenced very much by the skill of the decorator and the nature/complexity of the surface being decorated, a measure of the relative ease of lapping of different products can be obtained by controlled tests designed to pick up 'wet edges' after varying time intervals ranging from 2–20 minutes, assessing when dry for uniformity of sheen and freedom from unacceptable defects. Similarly, *recoatability* is best established by practical tests, recoating after different time intervals (depending on the type of product and its normal drying time). Solvent sensitivity tests may often be used as a guide to recoat times.

In application by roller, *spatter resistance* is becoming increasingly important. Ultimately, the best method of assessment is to roll a standard test area with a black card mounted either behind the roller on the roller handle, or below the test area. In carrying out comparative tests, it is necessary to control as many aspects as possible, such as the nature of the roller, the speed of rolling and the amount of pressure applied.

Surface defects such as inadequate or excessive flow, or sagging, are influenced by the speed of recovery of structure within the film, once spreading has ceased. They will also be influenced by film thickness and by the drying conditions and porosity of the substrate, making testing more of a comparative nature than an absolute measurement. Relative flow properties can be established by controlled drawdowns with wire-wound bar coaters on hiding power charts, whilst sagging tendency is normally assessed with serrated edge tapered gap doctor blades, applying on a standard non-absorbent substrate and standing the test panel vertically.

Colour uniformity tests may vary widely depending on the type of product and the characteristics of the surfaces to which it is normally applied. In gloss finishes the most likely defect is flotation or flooding leading normally to a deeper colour at the surface than in the depth of the coating. Increased flotation may occur where the film is thicker and may be uniform or striated or mottled. Since flocculation may be responsible, a common technique is to apply a blob of unsheared paint to the surface of a brush-applied panel, immediately after application. Another portion of the brush-applied film is rubbed with the tip of the finger in a circular motion when the film is beginning to set, to create additional shear and deflocculate the film as far as possible. When dry, the areas of high and low shear can be viewed for colour differences compared with the normally applied film. In the case of flat or sheen latex coatings, colour uniformity problems are more likely to occur as a result of varying degrees of coalescence or film porosity, caused by differences in the absorption of the substrate, or by significant variations in drying conditions between different applications. Typical of such problems is when new work is touched up several weeks later, to repair minor damage caused by other tradesmen during the finishing stages of new construction. So called 'patching tests' should simulate the extremes of ambient conditions likely to be experienced in practice. Reference has previously been made to the use of test substrates with varying absorption to confirm the products ability to give uniform colour over such surfaces.

8.3.2 *Industrial finishes*

A host of different finishing methods are used in the industrial market ranging from conventional, airless, hot, automatic or electrostatic spray, to flow-coating, conventional dip, electrophoretic deposition (which may be anodic or cathodic), roller coating or bar coating. Within any one of these categories, detailed but significant variations may exist within each finishing plant. In evaluating a coating for the acceptability of its application characteristics it is important to simulate as closely as possible the conditions applying at the customer's plant. As has been stated previously, the real test is the limited trial at the factory, but there are circumstances where this is not possible, particularly where large dip tanks are involved. In such cases the sheer scale of the problem in making a change from one product to another makes it all the more important for both supplier and customer to be satisfied that all is well before the change takes place. This comes back to the question of simulation and realism within the evaluation. As with the finished product, the testing schedule and methods used will tend to become customer specific. Within the evaluation, more objective tests may be used to establish the relationship between the product under test and an established product. Rheological comparisons are clearly important for all methods of finishing, while in the case of the various electro-painting techniques, information on electrical properties of the liquid coating, or deposited film is important, together with an assessment of 'throwing power' for electro-deposition primers. In most cases the customer should have a clear idea of the requirements, if not a full specification.

8.4 Film formation under application conditions

It is appropriate to consider this aspect under the headings of air drying solvent-borne coatings, air-dry latex coatings and cross-linking products. One of the key factors to consider is the *range* of ambient conditions likely to be experienced within the product's use. This range will vary widely from one country to another and from exterior applications to interior finishing in a variety of different atmospheres. The wide differences which exist can mean that products fully acceptable in some countries are quite unsuitable for others. Humidity and air movement can be just as important determining factors as temperature. Apart from comparative testing under controlled laboratory conditions, additional tests should always be carried out under simulated extreme conditions.

8.4.1 *Air-dry solvent-borne coatings*

Such coatings normally dry by a combination of oxidation and polymerisation, catalysed by the drier metals present. In a few cases the film formation is achieved purely by solvent evaporation, but then normally remains solvent sensitive. Reference has already been made in Section 8.1.1 to the problems encountered with the widely used BK drying recorder. Various other automated drying recorders exist using different means of assessing the degree of drying. Those employing a trail of sand or applying a strip of muslin or flock highlight the 'tack free' time but give less indication of other stages. 'Mechanical thumb' tests concentrate on the 'hard dry' stage – the loading used being appropriate to the type of product.

Solvent rub tests are commonly used to assess the recoat time for primers or undercoats. In order to overcome the unreliability of the standard BK drying test (using standard needles), the Paint Research Association studied a number of alternative methods. Of these, two were identified as giving statistically reproducible results and useful indicators of various stages of drying – both being based on the established BK recorder. One method used PTFE-coated ball-tipped needles, as used in the Gardner circular drying time recorder, in place of the standard BK needles. The second method involved replacing the normal arm with a special arm holding six small units which act as solvent reservoir/dispensers. As the arm traverses the length of the six adjacent test strips, at 1 inch intervals, an activator arm causes each unit to dispense one small drop of solvent on to each coating. Of various solvents tested, *p*-xylene gave the most consistent results. It was found that five distinct stages in drying could be observed:

- Stage 1. A circle with a clear centre and raised circumference.
- Stage 2. A wrinkled circle which lacks a defined circumference.
- Stage 3. A circle with a defined, non-raised circumference containing clear and wrinkled areas.
- Stage 4. A circle with a defined, non raised circumference containing a clear slightly patterned area.
- Stage 5. An unmarked surface.

Tests indicated that most films became touch dry between Stage 2 and Stage 3. Attempts have been made by the PRA to interpret the reasons for the distinct stages, but further correlation tests are required to relate these results to those obtained by more established drying recorders. A prototype Automated Solvent Droplet Tester has now been developed by Mickle Laboratory Engineering Co. Ltd.

8.4.2 *Air dry latex coatings*

Film formation from such coatings occurs in several stages. Water first evaporates or is drawn into the substrate allowing the resin particles (and pigment) to concentrate to the point where contact occurs. Aided by the coalescing solvent present, the particles coalesce to a continuous film. Finally the remainder of the water and coalescing solvent evaporate to leave the film at its maximum potential strength and block resistance, depending on composition. This describes the expected sequence under 'normal' drying conditions.

Several factors can upset this sequence, resulting in weak or improperly coalesced films – such abnormal conditions should be evaluated, if anticipated in practice. Possibly the most significant adverse condition during the drying process is high humidity, retarding the rate at which water is lost from the film, particularly if applied to a near saturated substrate. In such circumstances, it is possible to lose sufficient of the coalescing solvent to adversely affect film coalescence when the resin particles ultimately come into contact. Full coalescence can also be affected by low temperature at the paint surface during the fusion stage and also by too rapid a loss of water/solvent when applied to very absorbent substrates. Such potential defects can be overcome by formula modification, providing the conditions are anticipated and tests carried out to simulate them. In employing low temperature coalescence tests, the paint should be applied to an absorbent surface, precooled to the required temperature. Use of a non-absorbent temperature gradient bar can give misleading results.

8.4.3 *Cross-linking products*

The key to evaluation of such products is to ensure that the test conditions simulate as closely as possible the conditions likely to be experienced in practice. In many cases this is quite straightforward but, depending on the curing method being used, results may be influenced by the colour of the coating or by the nature/bulk of the substrate. Where reproduction of the anticipated curing conditions is not realistic within the laboratory, it is necessary to carry out correlation tests to ensure that results are predictable. Tests for the achievement of cure are discussed within the next section.

8.5 Film properties

It is worth repeating that whichever test or method is selected to evaluate one or other of a paint film's properties, attempts should be made to standardise conditions as much as possible to ensure the maximum degree of reproducibility. Aspects to consider are use of a standard carefully selected substrate, application of a uniform thickness film, curing under appropriate controlled conditions, ageing the film for an adequate length of time or under accelerated conditions to realise its ultimate strength before testing, and actually carrying out the test under as controlled conditions as possible. Instruments used to make measurements should be checked for accuracy by re-zeroing before use and by regular calibration tests. Ideally, conditions under which the test was carried out should be recorded, as well as the results. In view of the multiplicity of tests and methods already established it is useful to record variations in these methods for local or company usage within a company test methods manual, which should be updated whenever changes or additions are required. For convenience tests will be classified under the headings of film appearance, physical properties and chemical properties.

8.5.1 *Film appearance*

Many instruments are available today which can give a direct printout of the *whiteness*, *colour* or *colour difference from standard*, of a film. In the case of tristimulus colorimeters, results are expressed in terms of X, Y and Z values, whilst with modern spectrophotometers a plot is produced of reflectance at different wavelengths through the spectrum. In colour difference measurements results are computed in terms of CIE values of L, a, and b, which can be particularly useful for control purposes, enabling the ΔE values of individual batches against standard to be plotted. The actual method used for measurement or recording will be determined by the instrument available. The main general precautions to be taken are to ensure that an adequately thick film has been prepared to give full hiding, and that the film is properly dry before measurement.

Similarly, in measurement of *gloss or sheen*, numerous meters are available, with either variable geometry or with fixed heads measuring at 20°, 45°, 60° or 85° to the normal. The author prefers the use of fixed head instruments, using 20° meters for measurement of gloss finishes and 60° for low sheen finishes. For flat or very low sheen finishes, an 85° glossmeter is preferred. Various subjective methods of assessment rely on establishing a box with controlled angles of illumination and viewing, and assessing the sharpness of image of a reflected background which may be chequered, lined or have contrasting figures. Such techniques can be acceptable for direct comparisons and in some cases may give a more critical judgement than the glossmeter reading.

Measurement of *opacity* or *contrast ratio* (CR) can also range from very simple side-by-side drawdowns on hiding power charts to complex determination of spreading rate at 0.95 contrast ratio, which involves the measurement of percent reflectance over black and white of a number of films of different (but accurately measured) thicknesses, plotting the results and interpolating to 0.95 CR. Providing attention is paid to the attainment of uniform thickness wet films, using an applicator laying down a film approximately equal to normal spreading rates, direct measurement of percent reflectance over black and white may be adequate for many purposes. For detailed testing, one should refer to ASTM D2805, ISO 6504/1–1983, BS 3900 : Part D7 : 1983 or DIN 53162 as accepted international methods.

Yellowing tendency of alkyd-based films can be assessed by storage of films in a dark cupboard, whilst leaving the film exposed to air. Exclusion of natural air circulation can minimise yellowing, whilst presence of nitrogenous materials in the atmosphere will accelerate it. For this reason, the Woburn test places the films in a desiccator containing an ammonium salt, within the dark environment.

Blooming of alkyd films, producing a dull haze over the surface of the film, can be caused by the deposition of ammonium salts from the atmosphere. Tests to reproduce this require a similar atmosphere and a common method of test is to hang the test panels in a washroom or toilet.

8.5.2 *Physical properties*

The normally accepted test for *adhesion* of coatings is the *cross hatch* test. Results can be influenced very much by inadequate attention to detail, such as in panel preparation, or in how the cuts are made in the film or in the application or removal of the self-adhesive tape. For reproducible results, conditions should be standardised as much as possible, using preset parallel blades to make the cuts at right angles to one another. The percentage of squares left adhering to the substrate is a measure of the adhesion of the coating.

Bend tests, whilst normally being regarded as a measure of *flexibility* of the coating, also give an indication of any weakness in adhesion. Various gauge mandrels are specified for different coating systems, and the tests may be carried out at room temperature or at low temperature.

An alternative flexibility test is use of the *Erichsen tester*, in which a hemispherical indenter is forced into the panel from the reverse side, and the indentation depth at which the film starts to crack is a measure of its flexibility. Control of panel and coating thickness, temperature and deformation rate is important for reproducibility.

The *hardness* of the coating may be measured by the commonly used '*pencil scratch*' *hardness* test, or the *weighted scratch needle* drawn across the film at constant speed. An alternative measure of hardness is the *pendulum hardness* using either the Koenig and Persoz pendulum or the Sward rocker. Other tests which tend to measure a combination of film properties are the *falling weight* test, in which a weighted indenter falls from a fixed height onto the coated surface or the back of the panel (reverse impact), or the *pendulum test* in which coated rods are attached to swinging arms such that when the arms swing together the tubes strike and one (pivoted) tube slides across the other.

These tests have been established over many years, mainly for the testing of industrial finishes on metal substrates. The development of latex coatings in the decorative market has led to the introduction of a new range of tests more appropriate to such coatings. *Scrub resistance tests* are now established, using a number of alternative mechanical scrub machines, various brushes or pads, with water or detergent solution, and a number of techniques for measuring the speed or degree of wear or breakdown of the coating. If non-absorbent substrates are used, the test tends to give a measure of the *wet adhesion* of the coating, whilst testing over a more porous substrate gives a measure of the coating's *erosion resistance*. For comparative purposes, a measure of the relative erosion resistance may be obtained by rubbing across a film, drawn down on a hiding power chart, with a wet cloth noting the number of rubs to first point of removal of the coating and the number required to remove a complete line of the film. A similar test may be carried out on a number of films applied side by side on a larger panel, using a nail brush and water or detergent solution.

For low pigment volume concentration (PVC) films, testing for *block resistance* gives an indication of the film's thermoplasticity and tendency to block when in contact with a similar film, under pressure. Such testing is important for coatings likely to be applied to windows. A typical method would be to apply films to microscope slides and after various drying times to place two slides at right angles to one another, with the painted films in contact under a fixed load. After a set interval, the force required to separate the slides is determined. A separation force less than the applied force is normally an indication of an acceptable result.

Film *porosity* can be measured in a number of ways, the choice of method being determined by the type of coating and the influence porosity may have on its general performance. In a flat latex coating, increasing porosity will reduce its stain resistance and stain removal characteristics. *Stain resistance* tests such as the 'gilsonite' test, or more specific tests using typical stains, e.g. lipstick, crayon, pencil, ball-tip pens, etc., are appropriate, using a variety of cleaning solutions to establish ease of removal. In products which may be overcoated with decorative gloss finishes, *enamel hold out* tests would be more appropriate, using gloss measurements to assess the degree of sinkage into the substrate. The test may be made more sensitive by overcoating with a semigloss enamel. In the case of products designed for application to exterior wood, *film permeability* to water vapour is increasingly recognised as an important property of the coating. In measuring permeability, it should be recognised that the rate of passage of vapour is inversely proportional to film thickness. A common technique for such measurement is use of the Payne cup, in which films of known thicknesses are sealed into the rim of shallow aluminium dishes which contain a desiccant, the films acting as a permeable membrane. The sealed cups are placed in a controlled atmosphere with high humidity, and the increase in weight of the cups measured at intervals over a period of time. Results can be plotted to show the rate of transmission of water against film thickness. From such graphs one can establish the 'specific permeability' of a coating, at unit thickness or the 'film permeability' which relates to its average film thickness, in practice. Units of measurement have varied for different workers, but typically would be $g/m^2/24$ h (0.1 mm film thickness). The temperature and humidity gradient across the film should also be recorded to avoid comparing results obtained under different conditions.

Alternative, more practical tests may be carried out for comparative purposes, such as fully coating a section of joinery timber with the system to be tested, well sealing the end grain, and submitting it to a cycle of exposure to high humidity, low temperature, and incubation at 40°C, weighing after each exposure to note weight gain or loss. Such tests can also be used to assess the *blister resistance* of the coating or system. The degree of blistering can be assessed by comparison with photographic standards published in ASTM Standard Method D714-56.

8.5.3 *Chemical properties*

Chemical resistance tests apply more specifically to industrial finishes and are thus normally carried out on films applied to adequately prepared metal panels. According to the product being tested, the panel may be steel, aluminium or tinplate but in all cases the panel should be thoroughly degreased, finely abraded, and where appropriate pretreated, taking care to avoid rehandling after preparation. Primers or topcoats are normally tested as part of a system, using an approved non-experimental product as the remaining part of the system.

In *water-resistance tests*, the panels are immersed in a thermostatically controlled bath of distilled water (38°C is commonly specified) and the panels are examined at 24-hour intervals for blistering and loss of gloss or whitening, after wiping carefully with a soft dry cloth.

An alternative method is to place the panels in a humidity cabinet, in which relative humidity is maintained at 100% and the temperature cycles between 42° and 48°C on an hourly basis, producing constant condensation on the face of the panels. Again, panels are inspected at 24-hour intervals.

Salt spray resistance is probably the most common test used for assessing *corrosion resistance*, and in this case the paint coating is normally cut, either vertically or diagonally, from top to bottom corners, using a scalpel, ensuring that the cut penetrates the film and scores the metal surface, before being placed in the salt spray cabinet. The salt solution used is prepared to the composition specified, to ensure continuity of results. Panels are normally examined after 48 hours and 1, 2 and 3 weeks for signs of blistering, loss of adhesion and corrosion, particularly around the area of the cut. In some specifications, an intermittent salt spray is used, the mist-forming spray operating for only 10 minutes in every hour.

Alkali and detergent resistance tests are often specified, particularly for domestic appliances, and commonly use solutions of trisodium orthophosphate, soda ash (anhydrous sodium carbonate) or sodium hydroxide, giving differing degrees of aggressiveness. Panels are normally only partially immersed to facilitate assessment of the effect of the alkali, and are inspected after 4 hours and at 24-hour intervals, washing off with clean water and drying before assessment. Detergent-resistance tests are carried out in a similar way, the choice of detergent usually specified or agreed with the customer, and the inspection usually carried out at more frequent intervals.

Solvent-resistance tests would normally apply to automotive finishes, or finishes likely to come into contact with petrol (gasoline) or diesel fuel, whilst the full cure of a coating may be assessed by immersion in a polar solvent such as MIBK (methyl isobutyl ketone). Inspection of panels is made after 1, 2, 4, 6, 24 and 48 hours. A solvent rub test may sometimes be used as an alternative to partial immersion.

8.5.4 *Durability testing*

Durability of coatings may be tested by natural exposure, i.e. by placing panels normally at 45%, facing due south (in the northern hemisphere), or by accelerated weathering tests, using one of a variety of commercial weatherometers. The latter employ various cycles of exposure to water spray and UV light to simulate normal climatic changes.

Attempts have been made by many workers to correlate results of accelerated and natural weathering, with varying degrees of success (or failure). One of the problems in many countries is the variability of the natural weather conditions, which makes any attempt at correlation impossible. However, providing this is recognised, accelerated tests have the benefit of providing standard reproducible conditions, and, particularly if a control panel of known performance is included in the test series, can be of value in assessing the likely relative performance of a series of paints.

As with other forms of testing, selection of an appropriate test substrate is important, including its preparation. In the case of finishes designed for use on metal, steel or aluminium panels are normally used. Masonry finishes were commonly tested on asbestos cement panels, but since the concern on the use of asbestos, alternative flat masonry type panels are preferred. However, since few of these are as alkaline as traditional asbestos cement and have different levels of porosity, care must be taken in comparing results on the two alternative substrates. Testing of wood finishes introduces a number of problems which must be taken into account if results are to be valid. Panels should be carefully selected from, if possible, the same length of timber. Any sections showing abnormal defects or excess porosity should be rejected. The wood should be of a similar type to that used in practice. When paints for joinery are being tested, typical cross section timber should be used, rather than flat panels, incorporating a simulated joint or water trap to reproduce the typical area where failure normally starts.

Panels are assessed for performance at regular intervals, including gloss, colour and dirt retention, mould growth, chalking, checking and cracking, blistering, flaking and rusting, wherever appropriate. Normally an arbitrary 0–10 scale is used, and the results are plotted on an exposure record chart, which includes details of the system applied, date of exposure, details of the substrate used, the purpose of the exposure and reference to other panels in the series. Such examinations may typically be made at monthly intervals up to 4 months, at 2-monthly intervals from 4 to 12 months, at 3-monthly intervals in year 2 and thereafter at 4-monthly intervals until withdrawn from exposure. It is also common practice to wash the same half of the panel at each inspection.

Appendices

Appendix I Paint calculations

Chapter 6 indicated availability and use of simple computer programs which can take all of the tedium out of conversion of formulae to alternative batch size or units, and the calculation of basic constants, such as solids or volatile content, specific gravity, PVC, etc. Despite the availability of such programs, the basics of such calculations should be understood, allowing use of an electronic calculator to achieve the same results. There are times when a simple slide rule will provide the answer, to the required degree of accuracy, far quicker than either of the above methods.

Weight/volume conversions

The most basic calculation required is the conversion of quantities of raw materials or total formulations from weight to volume or vice versa. While this should present no problem to anyone engaged in paint development, confusion can arise due to the various units used in different literature or markets. Specific gravity (SG) is the most common factor and its use is recommended at all times. Use of the term lb/gal should be viewed with caution, since the gallon measure is rarely specified as US or Imperial. However, it must be recognised that many US raw material suppliers offer recommended formulations to a yield of 100 gallons (US) and quote density of raw material in lb/US gallon. In such cases, the lb/gal conversion factor must be used, or the formulation first converted to a total yield of 1000 litre, and specific gravity used. Basic relationships are given in Table A.1

Whilst raw material suppliers normally quote either specific gravity or density for their material, as supplied, in the case of solutions or latices they do not always quote a specific gravity for the solid resin. This figure is required for calculation of binder volume relationships within the formulation, and can easily be calculated providing the specific gravity of the volatile component is known. Examples of such calculations for a long oil alkyd and a typical latex are given in Table A.2.

Table A.1 *Weight/volume conversions*

Specific gravity (SG)	=	$\frac{\text{Weight (g)}}{\text{Volume (ml)}}$ or $\frac{\text{(kg)}}{\text{(l)}}$ (water = 1.0)
Hence: volume (l)	=	$\frac{\text{Weight (kg)}}{\text{SG}}$
or weight (kg)	=	Volume (l) × SG
Using American convention:		
Weight/gal (density)	=	$\frac{\text{Weight (lb)}}{\text{Volume (gal)}}$ (water = 8.33 lb/gallon)
Hence: volume (US gallon)	=	$\frac{\text{Weight (lb)}}{\text{Wt/gallon}}$
Bulking value is the reciprocal of weight/gal		
Hence: volume (US gallon)	=	Weight (lb) × bulking value (gallon/lb)
To convert from one convention to another:		
Specific gravity	=	lb/US gallon× 0.12
lb/US gallon	=	SG × 8.33

Table A.2 Specific gravity of resin solids

(a) Long oil alkyd	Given:		Resin solids = 70% in white spirit SG of resin solution = 0.97 SG of white spirit = 0.79		
100 kg resin solution		=	$\frac{100}{0.97}$	=	103.1 l
30 kg white spirit		=	$\frac{30}{0.79}$	=	38.0 l
∴ 70 kg resin solids		=	(103.1− 38.0)	=	65.1 l
i.e. SG of resin solids		=	$\frac{70}{65.1}$	=	1.075
(b) 100% acrylic latex	Given:		Resin solids = 53.5% by weight SG of latex = 1.044 SG of water = 1.00		
100 kg latex =	$\frac{100}{1.044}$			=	95.79 l
46.50 kg water				=	46.50 l
∴ 53.50 kg resin solid		=	(95.79 − 46.50)	=	49.29 l
i.e. SG of resin solid		=	$\frac{53.5}{49.29}$	=	1.085

Formulation calculations

The optimum level of addition of minor constituents is often calculated as a given percentage, based on the total formulation or on one of the key major ingredients. Such calculations should present no problems. An example of a slightly more complex calculation is the addition of metal driers to air drying (long oil) alkyds, where convention suggests the use of quantities such as 0.5% lead (now being replaced by less toxic materials), 0.2% calcium or 0.05% cobalt, as metal, calculated on the resin solids. Suppliers of drier solutions normally specify the weight percentage of metal present, e.g. 24% lead naphthenate, 10% calcium, 6% cobalt octoate, etc. An example of how to calculate the amount of drier solution is given in Table A.3.

Another common calculation is the determination of the ratio in which to blend two raw materials of different total solids content, to achieve a required

Table A.3 Calculation of drier addition

Given: Alkyd resin at 75% solids in white spirit
1000 l formulation contains 618 kg resin solution
Lead drier – 24% lead metal in solution
Calcium drier – 10% calcium metal in solution
Cobalt drier – 6% cobalt metal in solution

Weight of resin solids in 1000 litres	$= \frac{618 \times 75}{100}$	= 463.5 kg
Required amount of lead metal at 0.5% metal on resin solids	$= \frac{463.5 \times 0.5}{100}$	= 2.3175 kg
∴ Required amount of lead drier solution	$= \frac{2.3175 \times 100}{24}$	= 9.66 kg

Similarly, for a calcium metal content of 0.2% on resin solids

we require $\frac{463.5 \times 0.2}{100} \times \frac{100}{10}$ = 9.27 kg calcium drier

and for cobalt at 0.05% metal on resin solids

we require $\frac{463.5 \times 0.5}{100} \times \frac{100}{6}$ = 3.86 kg cobalt drier

Table A.4 Blending of two components to achieve a required solids content

% in solution A → % needed → *Quantity* solution A
% in solution B → % needed → *Quantity* solution B

Example 1: consider blending of solution A @ 85% solids and solution B @ 37% solids to give a final 60% solids.

85%			23	47.9 solution A
	60%			
37%			25	52.1 solution B
			48	100

The quantities of solutions A and B are established by noting the difference between their solids contents and the required percentage (following the lines indicated). Percent composition is then simply calculated.

As a check on the figures:

85% of 47.9	=	40.7	Solids in solution A
37% of 52.1	=	19.3	Solids in solution B
100.0		60.0	Total solids in mixture

Example 2: to reduce a composition from 55% to 52% by addition of thinner.

55		52	94.5
	52		
0		3	5.5
		55	100.00

i.e. 3 parts of thinner are added to 52 parts of the original composition, giving an additional 5.5% thinner in the final composition final solids content. A similar approach is taken, whether blending two solutions, two finished products or adjusting to a lower solids content by addition of solvent. The basic approach is shown in Table A.4.

Conversion of formulations to alternative units

Throughout this book, the value of formulating on the basis of a yield of 1000 litres has been stressed. However, many formulations used as a basis for development of new formulations may be found to be calculated on a different basis, particularly to a yield of 100 US gallons or on a weight basis to a total yield of 100 parts by weight (pbw). Further complications may arise if dry ingredients are entered by weight and liquid ingredients by volume. Table A.5 summarises the steps necessary to convert from one basis to another.

Table A.5 *Conversion of formulae to alternative units and yields*

Total yield		
Existing	Required	
1. 100 pbw	1000 litres	Calculate the SG of the existing formulation. Multiply the parts by weight of individual raw materials by 10 × formulation SG to give kilo/1000 litres
2. 100 US gallons	1000 litres	Multiply the weight in pounds of individual raw materials by 1.2 to give kilo/1000 litres. Multiply the volume in gallons of raw materials by 10 and enter as litres
3. 100 gallons (Imperial)	1000 litres	Change pounds to kilos and gallons × 10 to litres
4. 1000 litres	100 US gallons	Multiply the kilo weight of individual RMs by 0.833 and enter as pounds. Divide any litre quantity by 10 and enter as US gallons
5. 1000 litres	100 (Imperial) gallons	Change kilo to pounds and divide litres by 10 to enter as Imperial gallons
6. 100 US gallons	100 (Imperial) gallons	Multiply all pounds figures by 1.2, but leave gallon figures unchanged
7. 100 gallons (Imperial)	100 (US) gallons	Multiply all pounds figures by 0.833, but leave gallon figures unchanged
8. 100 pbw	100 US gallons	Calculate the formulation SG. Multiply the parts by weight of individual raw materials by the formulation SG × 8.33 to give weight in pounds

Calculation of paint constants

Reference was made in Chapter 1, Section 1.5 to the value of using a formulation breakdown sheet (see Figure 1.4), particularly to assist in calculation of paint constants. Examples of completed calculations for a long oil alkyd-based gloss finish and a typical exterior latex based formulation are given in Figures A.1 and A.2.

PRODUCT: Air dry gloss enamel
COLOUR: White

CODE NO: 632-66
FORMULATION DATE: 25.10.90

STANDARD/ACTUAL COSTING
DATE:

(A) Specific gravity	RAW MATERIAL	(B) KILO	(C) LITRE		(D) WEIGHT SOLID	(E) WEIGHT VOLATILE	(F) VOLUME SOLID	(G) VOLUME VOLATILE	+ %	GROSS USAGE	STD/ACT PRICE	COST	TOTAL COST
4.0	Titanium dioxide	330.00		PIGMENT	330.00		82.50						
				EXTENDER									
0.97	Long oil alkyd (75%)	586.85	(605.00)	RESIN & OIL	440.14	146.71	419.29	185.71					
1.05	Soya lecithin	1.80			1.80		1.71						
0.84	Dipentine	16.80		SOLVENT		16.80		20.00					
0.79	White spirit	212.29				212.29		268.72					
1.8	Bentone 38	1.80		ADDITIVE	1.80		1.00						
0.97	1% Silicone DC200/20	2.70	(2.78)		0.03	2.67	0.03	2.75					
0.92	Methyl ethyl ketoxime	2.30				2.30		2.50					
0.97	10% Cobalt drier	2.20											
0.97	10% Calcium drier	8.80	(15.79)		11.12	4.77	9.75	6.04					
1.10	18% Zirconium drier	4.89											
				STAINER/ TINTER									
	Total	1170.43		TOTAL:-	784.89	385.54	514.28	485.72					

VISCOSITY:-		
BS Cup:- Sec	GROSS YIELD:-	1000
Krebs:- K.U.	LOSS ALLOWANCE:-	
Cone & Plate: Poise	NETT YIELD:-	
Rotothinner: Poise	METHOD OF MANUFACTURE:-	
2.5 poise (ICI)	Bead mill	

% SOLIDS (Wt) 67.1% (Vol) 51.4%
S.G.:- 1.17
P.V.C.:- 16%
P/b ratio:- 0.75:1

TOTAL COST:-
GROSS COST/LITRE:-
NETT COST/LITRE
TINT COST/LITRE
STANDARD/ACTUAL COST/LITRE:-

Figure A.1

PRODUCT: Exterior latex paint
COLOUR: White

CODE NO: XW-829-8
FORMULATION DATE: 25.10.90

STANDARD/ACTUAL COSTING
DATE:

(A) Specific gravity	RAW MATERIAL	(B) KILO	(C) LITRE		(D) WEIGHT SOLID	(E) WEIGHT VOLATILE	(F) VOLUME SOLID	(G) VOLUME VOLATILE	+ %	GROSS USAGE	STD/ACT PRICE	COST	TOTAL COST
4.10	Titanium dioxide	216.00		PIGMENT	216.00		52.60						
2.60	Minex 4	330.00		EXTENDER	330.00		126.80						
1.06	Rhoplex AC829 (55%)	342.48	(322.60)	RESIN & OIL	188.36	154.12	168.48	154.12					
1.11	Ethylene glycol	21.60		SOLVENT		21.60		19.40					
0.95	Texanol	14.16				14.16		14.90					
1.00	Water	428.00				428.00		428.00					
1.15	Dispersant (Tamol 963-35%)	11.04	(9.60)	ADDITIVE	3.86	7.18	2.42	7.18					
0.92	Defoamer (Colloid 643)	7.20			7.20		7.80						
1.36	Colloid (Natrosol 250MHR)	3.00			3.00		2.20						
0.97	Preservative (Super AD IT)	7.20			7.20		7.40						
2.40	Struct. agent (Attagel 50)	6.00			6.00		2.50						
1.09	Wetting agent (Triton X405)	3.60			3.60		3.30						
0.91	Ammonia (28%)	2.64				2.64		2.90					
				STAINER/ TINTER									
	Total	1392.92		TOTAL:-	765.22	627.70	373.50	626.50					

VISCOSITY:-
BS Cup:- Sec
Krebs:- 90-95 K.U.
Cone & Plate: Poise
Rotothinner: Poise
1.0 - 1.2p (ICI)

GROSS YIELD:- 1000
LOSS ALLOWANCE:-
NETT YIELD:-

METHOD OF MANUFACTURE:-
High speed dispersion

% SOLIDS (Wt) 54.9% (Vol) 37.35%
S.G.:- 1.393
P.V.C.:- 51.6%
P/b ratio:-

TOTAL COST:-
GROSS COST/LITRE:-
NETT COST/LITRE
TINT COST/LITRE
STANDARD/ACTUAL COST/LITRE:-

Figure A.2

The stages involved in completion of such tabulations are as follows:

1. Convert the formulation to 1000 litre yield, using procedures outlined in Table A.5, and enter weight composition (kilo) in column B.
2. Establish specific gravities (SG) of all ingredients and enter in column A.
3. Calculate volume of each ingredient by dividing each weight by appropriate specific gravity. If material is 100% solid, re-enter weight in column D and volume in column F. For 100% volatile ingredients, enter weight and volumes in columns E and G respectively. Where material is a mixture of solids and volatile, enter volume in column C.
4. For such materials, calculate weight of solid material by multiplying total weight of each ingredient by its respective solids content. Actual weight of volatile is obtained by difference. Enter solid weights in column D and volatile weights in column E (i.e. D + E = B).
5. Convert the quantities of volatile to volume by dividing by the specific gravity of the solvent/diluent (e.g. 0.79 for white spirit or 1.0 for water) and enter in column G.
6. The volume of resin/solid material can be calculated by subtracting the volume of volatile (column G) from the total previously entered in column C, and entering it in column F.
7. Calculate totals for columns B, D, E, F and G.
   Note: Total of D + E = Total B
   Total of F + G = 1000
8. Calculate percent total solids by weight and volume as follows:

$$\text{Wt. solids} = \frac{\text{Total column D} \times 100}{\text{Total column B}}$$

$$\text{Vol. solids} = \frac{\text{Total column F}}{10}$$

9. $$\text{Specific gravity} = \frac{\text{Column B}}{1000}$$

10. $$\text{PVC} = \frac{\text{Total volume pigment} + \text{extender} \times 100}{\text{Total volume of pigment} + \text{extender} + \text{resin solids}}$$

11. $$\frac{\text{Pigment binder ratio}}{\text{(p/b ratio)}} = \frac{\text{Weight pigment}}{\text{Weight of solid resin}} \text{ expressed as a ratio}$$

Application/spreading rates

Having established the percentage volume solids by the method described in paragraph 8, it is a simple matter to establish theoretical relationships between dry film thickness and spreading rates by use of the equations given in Table A.6.

Table A.6 Application/spreading rates versus film thickness

1 *Dry film thickness* (μm) $= \dfrac{\%\ \text{Volume solids} \times 10}{\text{Spreading rate}(\text{m}^2/\text{l})}$

e.g. given: % vol. solids = 51% and spreading rate = $15\text{m}^2/\text{l}$

Dry film thickness $= \dfrac{51 \times 10}{15} = 34\ \mu\text{m}$

2 *Theoretical spreading rate* $= \dfrac{\%\text{Volume solids} \times 10}{\text{Dry film thickness } (\mu\text{m})} (\text{m}^2/\text{l})$

e.g. given: % vol. solids = 36% and required dry film thickness = 30 μm

Required spreading rate $= \dfrac{36 \times 10}{30} = 12\ \text{m}^2/\text{l}$

3 *Practical spreading rate*:

Spreading rate $(\text{m}^2/\text{l}) = \dfrac{\text{Wt of 1 l of paint (kg)}}{\text{Wt applied to } 1\text{m}^2 \text{ (kg)}}$

4 *Weight of wet paint required for test panel*:

Wet paint required (g) $= \dfrac{\text{Wt of 1 l (kg)} \times \text{Panel area } (\text{cm}^2)}{\text{Spreading rate } (\text{m}^2/\text{l}) \times 10}$

e.g. given: SG of paint = 1.17
Spreading rate = $15\ \text{m}^2/\text{l}$
Panel size = 20 cm × 30 cm

Then wet paint required $= \dfrac{1.17 \times 60}{15 \times 10} = 4.68$ g

Appendix II Units for the paint formulator

Compiled by Len Tysall (adapted with permission from Information Brief No. 4, Supplement to Progress Report No. 9, Autumn 1985, Paint Research Association)

Introduction

There is still some confusion between SI, CGS and British units. There are many lists of conversion factors between these, but such a list is more useful when coupled with a summary of the relevant definitions. This appendix aims to give an easy introduction to this rather complex subject. Inevitably this involves some risk of loss of precision; formal sources should therefore be consulted when necessary. The most convenient source is *Changing to the Metric System*, Anderton and Bigg (National Physical Laboratory), 5th edn (1979), HMSO. Those units and conversion factors judged to be most interesting in the coatings field have been selected.

Some background points

Associated with the development of the SI system of units, there has been a good deal of standardisation on related details of presentation. The main points are summarised below:

Presentation of numerical values

Exact relations between units are printed in heavy type. For example, 1 yard = **0.914 4** metre. This is exact. Where the number would involve numerous digits, it is presented correct to six significant figures – and in normal type. For example, 1 metre = 1.093 61 yards.

Numbers are presented with their digits grouped in threes forward or backward from the decimal point. Groups of three are separated by a space, not by commas. The decimal point on the line is favoured over the comma or the point positioned above the line. All this should be obvious from the 'right' and 'wrong' examples overleaf.

Right	Wrong
123 456	123,456
12 345.6	12 345·6
1 234.56	1234.56
123.456	123,456
12.345 6	12.3456
1.234 56	1.23456
0.123 456	0.123,456

Multiplying prefixes

Where the SI unit is inconveniently large or small a 'multiplying prefix' is used; for example, 'millimetre' (mm) if 'metre' (m) is too large. The relationship is, of course, always in powers of ten. The names of these prefixes, together with their symbols, are listed below:

Name	Symbol	Factor by which the unit is multiplied		
exa	E	10^{18}	=	1 000 000 000 000 000 000
peta	P	10^{15}	=	1 000 000 000 000 000
tera	T	10^{12}	=	1 000 000 000 000
giga	G	10^{9}	=	1 000 000 000
mega	M	10^{6}	=	1 000 000
kilo	k	10^{3}	=	1 000
hecto	h	10^{2}	=	100
deca	da	10^{1}	=	10
deci	d	10^{-1}	=	0.1
centi	c	10^{-2}	=	0.01
milli	m	10^{-3}	=	0.001
micro	μ	10^{-6}	=	0.000 001
nano	n	10^{-9}	=	0.000 000 001
pico	p	10^{-12}	=	0.000 000 000 001
femto	f	10^{-15}	=	0.000 000 000 000 001
atto	a	10^{-18}	=	0.000 000 000 000 000 001

Not all powers of ten are equally favoured – multiples of three are preferred. Strict adherence to this principle would rule out some very widely used units such as the centimetre (cm). Hence 'recommended' multiples are listed, favouring the rule of three but allowing some exceptions (see pages 234 and 235).

The position of the kilogram is anomalous. The name is so well established that it has not been possible to replace it by a 'prefix-free' alternative. Hence, there is some possibility of confusion; for example, lg (no prefix) is 10^3units of mass.

Miscellaneous points of style

(i) The name of a unit is written in lower case, even if it is derived from a proper name. For example, 'newton' – but, of course, (Sir Isaac) Newton.

(ii) The symbol of a unit, if derived from a proper name, is written with the first letter as a capital. For example 'pascal' (name) but 'Pa' (symbol). Sometimes this capital is the only letter. For example, 'newton' (name) but 'N' (symbol).

(iii) Symbols are not used in plural form. For example, '2 m' not '2 ms' for 2 metres. Note that symbols are *not* given a full stop to indicate that they are abbreviations. For example, '2 m' *not* '2 m.' – except, of course, at the end of a sentence.

Some definitions and related comments

Basis of the definitions

The SI system defines seven 'base units' and in principle all other units are derived from these. Definitions of these and of a selection of other SI units are summarised below and related to corresponding British and American units where appropriate. When first mentioned in the text, each unit is followed by its 'symbol' in parentheses (for a fuller list of symbols see pages 232 and 233). Exact values for definitions and conversion factors are given in heavy type. Where values are not exact, they are quoted correct to six significant figures in ordinary type. For a fuller set of conversion factors see pages 236 and 237.

The base units

Length The metre (m) is defined in spectral terms. The centimetre is a 'recommended' multiple in spite of being an exception to the 'rule of three'. The Ångstrom (Å) is giving way to the nanometre (nm) which is ten times as large. The use of the term 'micron' for µm is deprecated.

Mass The kilogram (kg) is defined in terms of an international prototype. The name tonne (t) is accepted as a 'special name' for the Mg.

Time The second (s) is defined in terms of a standard radiation period. The other familiar units – minute (m), hour (h), day (d) – also week, month and year – have their usual relationships.

Electric current If two straight parallel conductors carry a current of one ampere (A), there will be a force of 2×10^{-7} newtons between them for every metre of length (exact definition requires constant current, negligible cross-section of the conductors and vacuum).

Temperature The kelvin (K) is the fraction 1/273.16 of the thermodynamic temperature of the triple point of water. The degree Celsius is equal to the kelvin. However, the zero of the Celsius scale is *not* the triple point but the ice point (**273.15 K**). The term 'degree Centigrade' was renamed 'degree Celsius' in 1948 to avoid confusion with other names.

Definition of the Fahrenheit scale is somewhat uncertain, but it is taken to share the fixed points of the Celsius scale (0°C = 32°F; 100°C = 212°F). The well established conversions between the two scales are thus unchanged.

Amount of substance One mole (mol) contains the same number of 'elementary entities' (atoms, molecules, ions, electrons, etc.) as there are atoms in 0.012 kg of carbon 12.

Luminous intensity The candela (cd) is defined in terms of radiation from a defined area of black body at fixed temperature and pressure.

Other units

Volume Litre (l) is recognised as a special name for the cubic decimetre. Thus the old relationship 1 litre = 1.000 028 dm^3 no longer holds. Related to this 1 millilitre = **1** cm^3 (exactly).

The UK gallon is defined as being equal to **4.546 09** l. The US gallon is defined as **231** in^3 (=3.785 411 l). The 6:5 relationship between the two is no longer exact.

Density Density is expressed in the usual way in terms of weight /$(length)^3$ and no special names are needed.

Force The newton (N) is the force that will give an acceleration of 1 m/s^2 to a mass of 1 kg. Thus the dyne of the centimetre-gram-second (CGS) system (1 cm/s^2 to 1g) is equal to $\mathbf{10^{-5}}$ N.

Technical units of force such as the 'kilogram weight' and 'pound weight' are inexact because of variations in the Earth's gravitation. They can be given accurate values if related to the international standard acceleration of freefall (g_n = **9.806 65** m/s^2). Such units are correctly called 'kilogram-force (kgf)' and 'pound force (lbf)'. For example, one kilogram-force gives a mass of 1 kg an acceleration of g_n (the kilogram-force is called kilopond (k_p or kP) in some European countries). Such units are regarded as obsolescent (but see next section).

Pressure The pascal (Pa) is the pressure produced by a force of 1 newton applied, uniformly, over an area of 1 m^2.

'Pounds per square inch' (psi), meaning strictly pounds weight per square inch, is inexact (see previous section). Hence, conversion factors are given in terms of 'pounds force per square inch' (lbf/in^2). The relationship is 1 pound weight = 32.174 0 lbf, arising from the standard values of g_n (9.806 65 m/s^2 = 32.174 0 ft/s^2) .

There are three atmosphere units in use. The 'standard atmosphere' (atm) originally equal to 760 mmHg, is now defined as **101 325** Pa (equal to 14.695 9 lbf/in^2). However, the standard atmosphere is no longer used as a unit of pressure – the bar (bar) and millibar (mbar) are favoured (1 bar = **100 000** Pa; 1 mbar = **100** Pa). The 'technical atmosphere' (at) is also used. This involves the kilogram-force, sometimes called the 'kilopond' (kp). 1 kp/cm^3 approximates to one atmosphere and is therefore a convenient round number value (equal to **98 066.5** Pa, arising of course from the standard value of g_n. To summarise this rather complex situation:

1 standard atmosphere (atm) = **101 325** Pa = 14.695 9 lbf/in^2
1 bar = **100 000** Pa = 14.503 8 lbf/in^2
1 technical atmosphere (at) = **98 066.5** Pa = 14.223 3 lbf/in^2
(= 1 kp/cm^3)

Energy, work The joule (J) is the work done when the point of application of a force of 1 newton is displaced through a distance of 1 metre in the direction of the force.

The kilowatt hour (kW h) is accepted as a special name and is equal to **3.6** MJ, arising of course from definitions above. The old CGS 'erg' (or dyne centimetre) is thus equal to $\mathbf{10^{-7}}$ J .

Power The watt (W) is the power which gives rise to energy of 1 joule in 1 second.

Quantity of heat Definitions based on raising a standard mass of water through one degree of temperature lack precision because the energy required varies with temperature. For accurate work, therefore, temperature must be specified – for example, *not* 'calorie' but '15°C calorie'. But the use of the joule is favoured and therefore no 'recommended' multiples of the calorie are given.

There are two calories with exact definitions:-

(i) The international table calorie (cal) defined as **4.186 8** J.
(ii) The thermochemical calorie, defined as **4.184** J.

Other calories which are used include:

(iii) The 15°C calorie (cal_{15}) defined by reference to water from 14.5°C to 15.5°C, approximately equal to 4.185 5 J.
(iv) The 4°C calorie.
(v) The mean 0–100°C calorie.
(vi) The large calorie (equal to 1000 cal_{15})

Electrical units Electrical units other than the ampere, which is a base unit, are defined in traditional fashion. To summarise briefly:

(i) Coulomb (C) – quantity of electricity carried in 1 s by a current of 1A.
(ii) Volt (V) – difference in potential between two points of a conducting wire carrying a constant current of 1A when the power dissipated is 1 W.
(iii) Ohm (Ω) – electrical resistance between two points of a conductor when a constant potential difference of 1 V produces a current of 1 A.
(iv) Siemens (S) also known as the 'reciprocal ohm' or 'mho' – a conductance of 1 ampere per volt.
(v) Farad (F) – capacitance of a capacitor such that a potential difference of 1 V appears when it is charged by 1 coulomb.
(vi) Weber (Wb) – the magnetic flux which, linking a circuit of one turn, would produce an electromotive force of 1 volt if it were reduced to zero, uniformly, in 1 second.
(vii) Henry (H) – the inductance of a closed circuit in which an electromotive force of 1 volt is produced when the current varies uniformly at 1 ampere per second.

Viscosity The classical definition of dynamic viscosity defined the poise in terms of the force in dynes between parallel plates of 1 cm^2 area, 1 cm apart, moving with a relative velocity of 1 cm/s .Transposing this into SI units gives the pascal second (Pa s) which is ten times as large. The name 'poise' has been dropped to avoid confusion because the name 'poiseuille' has been used in France for the pascal second. But 'centipoise' (cP) is accepted as a special name for the millipascal second (mPa s) – the use of this name is likely to be discouraged.

In SI units kinematic viscosity is measured in m^2/s. 'Centistokes' (cSt) is accepted as a special name for the mm^2/s (note: centistokes *not* centistoke – the man's name was Stokes). Like the centipoise this name is likely to be discouraged.

Surface Tension The CGS dyn/cm is replaced by the N/m. The relationship is:

$$1\ \text{dyn/cm} = \mathbf{10^{-3}}\ \text{N/m} (= \mathbf{1}\ \text{mN/m})$$

Symbols for units

A number of these symbols have been given already in the text.

Symbol	Name of unit	Symbol	Name of unit
A	ampere	Btu	British thermal unit
Å	ångström		
at	technical atmosphere	C	coulomb
atm	standard atmosphere	°C	degree Celsius

Symbol	Name of unit
°C	degree Celsius
cal	I T calorie
cal_{15}	15°C calorie
cc	cm^3 abbreviation
cd	candela
cg	centigram
ch	cheval vapeur (h.p.)
cm	centimetre
cP	centipoise
cSt	centistokes
c/s	cycle/second
CV	cheval vapeur (h.p.)
cwt	hundredweight
d	day
dag	decagram
dg	decigram
dm	decimetre
dyn	dyne
erg	erg
F	farad
°F	degree Fahrenheit
fl oz	fluid ounce
ft	foot
ftH_2O	foot of water
g	gram
g_n	gravity, standard
gal	gallon
h	hour
H	henry
ha	hectare
hp	horsepower
Hz	hertz
in	inch
inHg	inch of mercury
inH_2O	inch of water
J	joule
K	kelvin
kcal	kilocalorie
kg	kilogram
kgf	kilogram-force
kJ	kilojoule
km	kilometre
kp	kilopond
kP	*see* kp
kPa	kilopascal
kW	kilowatt
l	litre
lb	pound
lbf	pound-force
m	metre
mb	mbar abbreviation
mbar	millibar
mg	milligram
Mg	megagram
mile	mile
min	minute (time)
ml	millilitre
mm	millimetre
mmHg	millimetre of mercury
mmH_2O	millimetre of water
MN	meganewton
mol	mole
mp	millipond
Mp	megapond
MP	*see* Mp
MPa	megapascal
μb	μbar abbreviation
μbar	microbar
μg	microgram
μHg	*see* μmHg
μin	micro-inch
μm	micrometre
μmHg	micrometre of mercury
N	newton
nm	nanometre
Ω	ohm
oz	ounce
P	poise
Pa	pascal
Pl	poiseuille
Ps	Pferdestärke (h.p.)
psi	'pounds per square inch'
pt	pint
qt	quart
s	second (time)
S	siemens
St	stokes
t	tonne
tf	tonne-force
ton	ton
tonf	ton-force
USgal	US gallon
V	volt
W	watt
Wb	weber
yd	yard
′	foot
′	minute (angle)
″	inch
″	second (angle)

Recommended multiples

Where the SI unit itself is inconveniently large or small a multiple should be used. Certain multiples are recommended for common use and are listed below, with their corresponding SI units. Some metric units with special names are accepted for use in conjunction with SI units – for example, litre. Those examples thought useful in the coatings field are quoted.

Quantity (SI unit with recommended multiples and sub-multiples)	Units with special names	
Plane angle		
rad mrad µrad	degree°	= π/ **180** rad
	minute′	= **1/60°**
	second″	= **1/60**
Length		
m km cm mm µm nm		
Area		
m^2 km^2 dm^2 cm^2 mm^2	hectare (ha)	= $\mathbf{10^4}$ m^2
Volume		
m^3 dm^3 cm^3 mm^3	hectolitre (hl)	= **0.1** m^3
	litre(l)	= **1** dm^3
	centilitre (cl)	= **10** cm^3
	millilitre (ml)	= **1** cm^3
Time		
s ks ms µs ns	day (d)	= **24** h
	hour (h)	= **60** min
	minute (min)	= **60** s
	also week, month and year	
Velocity		
m/s	km/h	= **1/3.6** m/s or 0.277 778 m/s
	m/h	
Frequency		
Hz THz GHz MHz kHz		
Rotational frequency		
s^{-1}	revolution per second (rev/s)	
	revolution per minute (rev/min)	
	reciprocal minute (min^{-1})	
Mass		
kg Mg g mg µg	tonne (t)	= **1** Mg
Density		
kg/m^3 Mg/m^3 kg/dm^3 g/cm^3	t/m^3	
	kg/l	= **1** kg/dm^3
	g/ml	= **1** g/cm^3
	g/l	= **1** kg/m^3

Quantity (SI unit with recommended multiples and sub-multiples)	Units with special names	
Force		
N MN kN mN μN		
Moment of force (torque)		
Nm MN m kN m mN m μN m		
Surface tension		
N/m mN/m (= mJ/ m^2)		
Pressure		
Pa GPa MPa kPa mPa μPa	kbar	= $\mathbf{10^8}$ Pa
	bar	= $\mathbf{10^5}$ Pa
	mbar	= **100** Pa
	μbar	= $\mathbf{10^{-1}}$ Pa
Stress		
Pa GPa MPa kPa	N/mm^2	= **1** MPa
Viscosity, dynamic		
Pa s mPa s	centipoise (cP)	= **1** mPa s
Viscosity, kinematic		
m^2/ s mm^2/ s	centistokes (cSt)	= **1** mm^2/ s
Energy, work		
J TJ GJ MJ kJ mJ	kilowatt hour (kW h)	= **3.6** MJ
Power		
W GW MW kW mW μW		
Heat, quantity of heat		
J TJ GJ MJ kJ mJ		
Heat flow rate		
W kW		
Specific heat capacity		
J/kg K kJ/kg K		
Specific energy, specific latent heat		
J/kg MJ/kg kJ/kg		

Selected conversion factors

Conversion factors which are judged likely to be useful in the coatings field, are given. To convert from units on right hand side to units on left hand side, multiply RHS units by value in column B. To convert the other way, multiply LHS units by value in column A. (Exact conversions in heavy type; other conversions are correct to six significant figures.)

	A	B	
Mass			
tonne (t) (= **1000** kg)	0.984 207	1.016 05	ton (ton) (= **2240** lb)
tonne (t)	1.102 31	0.907 185	US ton (= **2000** lb)
tonne (t)	2 204.62	–	pound (lb)

	A	B	
kilogram (kg)	2.204 62	**0.453 592 37***	pound (lb)
gram (g)	0.035 274 0	28.349 5	ounce (oz) (= **0.062 5** lb)
Length			
kilometre (km)	0.621 371	1.609 34	mile (mile)
metre (m)	1.093 61	**0.914 4**	yard (yd)
metre (m)	3.280 84	**0.304 8**	foot (ft)
metre (m)	39.370 1	**0.025 4**	inch (in)
centimetre (cm)	0.393 701	**2.54**	inch (in)
millimetre (mm)	0.039 370 1	**25.4**	inch (in)
millimetre (mm)	39.370 1	**0.025 4**	milli-inch (mil) ('thou') (= **0.001** in)
micrometre (μm)	0.039 370 1	**25.4**	milli inch (mil) ('thou')
Area			
km^2	0.386 102	2.589 99	$mile^2$
hectare (ha) (= **10,000** m^2)	2.471 05	0.404 686	acre (= **4840** yd^2)
m^2	1.195 99	0.836 127	yd^2
m^2	10.763 9	0.092 903 0	ft^2
cm^2	0.155 000	**6.451 6**	in^2
Volume			
m^3	1.307 95	0.764 555	yd^3
m^3	35.314 7	0.028 316 8	ft^3
m^3	6.289 82	0.158 987	US barrel (petroleum)
cm^3	0.061 023 7	16.387 1	in^3
litre (l) (= **1** dm^3)	0.219 969	**4.546 09**	UK gallon (gal)
l	0.264 172	3.785 41	US gallon (= **231** in^3)
l	1.759 75	0.568 261	pint (= **1/8** gal)
ml	0.035 195 1	28.413 1	fluid ounce (fl oz) (= **1/20** pint)
Velocity			
km/h	0.621 371	1.609 34	mile/h (mph)
m/s	3.280 84	**0.304 8**	ft/s
Acceleration			
m/s^2	3.280 84	**0.304 8**	ft/s^2
Density			
kg/m^3 (=g/l)	0.062 428 0	16.018 5	lb/ft^3
g/ml	0.016 018 5	62.428 0	lb/ft^3
g/ml	0.099 776 3	10.022 4	lb/gal (pounds per gallon)
Concentration			
g/l	0.160 359	6.236 02	oz/gal
g/cm^3 (=kg/l)	10.022 4	0.099 776	lb/gal
Force			
tonne-force (tf) (= **9.806 65** kN)	0.984 207	1.016 05	ton-force (ton f)
kN	0.100 361	9.964 02	ton f
N	0.224 809	4.448 22	pound force (lbf)
dyne (dyn)	$0.224\ 809 \times 10^{-5}$	$4.448\ 22 \times 10^{5}$	lbf
Pressure			
1 bar (bar) (=$\mathbf{10^5}$ pascal)	14.503 8	0.068 947 6	lbf/in^2 (psi)
1 millibar (mbar) (= **100** Pa)	0.014 503 8	68.947 6	lbf/in^2
standard atmosphere (atm)	14.695 9	0.068 046 1	lbf/in^2

*Note: Eight digits but exact

	A	B	
technical atmosphere (at) (= **1** kgf/cm^2)	14.223 3	0.070 307 1	lbf/in^2
1mm Hg (= 1 torr) (= 133.322 Pa)	0.019 336 8	51.714 9	lbf/in^2
Energy/work/heat			
joule (J)	0.737 562	1.355 82	ft lb f
MJ	0.277 778	**3.6**	kilowatt hour (kWh)
MJ	0.372 506	2.684 52	horsepower hour (hph)
International table calorie (cal_{IT})	0.003 968 32	251. 996	British thermal unit (Btu)
J	0.238 846	**4.186 8**	International table calorie (cal_{IT})
J	0.239 006	**4.184**	thermochemical calorie
J	0.238 92	4.185 5	15°C calorie (cal_{15})
J	**10^7**	**10^{-7}**	erg (erg)
Power			
Watt (W)	0.001 341 02	745.700	horsepower (hp) (= **550** ft lbf/s)
W	0.001 359 62	735.499	metric horsepower (ch,P s)
W	0.737 561	1.355 82	ft lbf/s
Viscosity, dynamic			
Pascal second (Pa s)	**1000**	**0.001**	Centipoise (cP) = mPa s
Pa s	**10**	**0.1**	poise (P)
Viscosity, kinematic			
m^2/s	**10^6**	**10^{-6}**	centistokes (cSt)
m^2/s	10.763 9	0.092 903 1	ft^2/s

Appendix III International test methods

Tests on liquid paints (excluding chemical tests)	BS 3900	ISO	ASTM
Sampling (liquid paints and related pigmented coatings)	Part A1:1970(1984)	= 1512:1974	D3925-81(1985)
Preparation of samples for testing	Part A2:1983(1985)	≡ 1513:1980	
Fineness of grind	Part C6:1983(1989)	≡ 1524:1983	D1210-79(1988)
Coarse particles in paints			D185-84
Package stability of paints			D1849-80(1987)
Settling of paint, evaluating degree of			D869-85
Density of paint	Part A12:1975(1984)	= 2811:1974	D1475-85
Freeze thaw resistance of water-borne paints			D2243–82(1987)
Viscosity measurement – by flow cups (by Ford Viscosity cup)	Part A6:1986	≡ 2431:1984	D1200-82
– by dip type cups			D4212-82
– using Stormer Viscometer			D562-81(1985)
– by Rotational (Brookfield) Viscometer			D2196- 86
– at high shear (using ICI Cone/Plate Viscometer)	Part A7:1986	≠ 2884:1974	D4287-88
Test for flash/no flash– closed cup equilibrium method	Part A8:1986	≡ 1516:1981	
– rapid equilibrium method	Part A13:1986	≡ 3680:1983	
Flashpoint determination – closed cup equilibrium method	Part A9:1986	≡ 1523:1983	
– rapid equilibrium method	Part A14:1986	≡ 3679:1983	
Combustibility – small scale test	Part A11:1974(1979)		

Chemical examination of liquid paints and dried films	BS 3900	ISO	ASTM
Water content, determination of – by Dean and Stark Method	Part B1:1965(1985)		
– by Karl Fischer Method			D4017-88
Volatile/non-volatile content – determination of	Part B2:1970(1984)	= 1515:1973	D2369-87
– volume non volatile	Part A10:1985	≡ 3233:1984	D2697-86
– volatile organic content (VOC)			D3960-87
Pigment content of paints – by low temperature ashing			D4451-85
Pigment content of water-reducible paints – by low temperature ashing			D3723-84
Pigment content of solvent-reducible paints – by high speed centrifuge			D2698-73(1984)

Acid extracts from liquid paints – preparation of	Part B5:1986	≡ 6713:1984	
Antimony, soluble – determination of	Part B7:1986	≡ 3856/2:1984	
Arsenic in paint – determination of			D2348-85
Barium, soluble – determination of	Part B8:1986	≡ 3856/3:1984	
Cadmium, soluble – determination of	Part B9:1986	≡ 3856/4:1984	
Chromium – hexavalent content of solid matter	Part B10:1986	≡ 3856/5:1984	
– total chromium content	Part B11:1986	≡ 3856/6:1984	D3718-85
Lead – soluble, determination of in solid matter in liquid paints	Part B3:1983(1989)		
– total lead in paints	Part B4:1986	≡ 6503:1984	
– soluble lead content	Part B6:1986	≡ 3856/1:1984	
– rapid method for estimation in liquid paint	Part B15:1987		
– in paint and dried paint films – detection of			D3618-85
– in paint by direct aspiration atomic adsorption spectroscopy (AAS)			D4834-88
Lead, cadmium and cobalt in paints by AAS			D3335-85
Mercury, – soluble, determination of	Part B12:1986	≡ 3856/7:1984	
– in paints, low concentrations of, by AAS			D3624-85
– total, flameless AAS		7252:1984	
Cellulose nitrate in alkyd lacquers by infrared (IR) spectrophotometry			D3133-72
Dichloromethane and 1,1,1-trichloroethane – analysis for			D4457-85
Vacuum distillation of solvents from solvent-reducible paints for analysis			D3272-76(1988)
Vehicle solids from solvent reducible paints – IR identification			D2621-87
Polymers in emulsion paints – qualitative identification of			D3168-85
Vehicle from solvent-reducible paints – separation of vehicle			D2372-85
Thinner in solvent-reducible paints – determination of nature			D2349-72(1984)
Titanium dioxide content in paint by X-ray fluorescent spectroscopy			D4764-88
Titanium dioxide content of pigments recovered from whole paint by AAS			D4563-86
Toluene di-isocyanate, free in urethane prepolymers and coating solutions			D3432-80

Application characteristics and film formation	BS 3900	ISO	ASTM
Brushing test – large scale	Part A5:1968	≠ TR3172:1974	
Natural spreading rate – determination by brush application	Part A16:1986	≡ 7254:1984	
Coating of test panels at specified spreading rate – by brush	Part A17:1986	≡ 7877:1984	D823-87
Thickness, wet film – measurement of			D1212-85
– by notch gauges			D4414-84
Dry film thickness – determination of	Part C5:1975(1987)	≡ 2808:1974	
– of protective coatings system (destructive measurement)			D4138-88
– measurement on ferrous base			D1186-87
– measurement on non-ferrous metal base			D1400-87
– by micrometer measurement			D1005-84
– on wood products by microscopical measurement			D2691-88
– of pipeline coatings on steel (non destructive)			G12-83
Flow ratings of organic coatings using Shell flow comparator			D2353-83
Leveling of paints by draw-down method			D4062-88
Sag resistance of paints using a multinotch applicator			D4400-84
Spatter resistance to roller application – measurement of			D4707-87
Low temperature coalescence of latex paint films			D3793-84
Wet edge time	Part C1:1965(1987)		
Drying times – surface dry (ballotini method)	Part C2:1971(1983)	= 1517:1973	
– hard dry time	Part C3:1971(1989)		
– of organic coatings at room temperature			D1640-83
Pressure test for stackability	Part C7:1982(1985)	≡ 4622:1980	
Block resistance of organic coatings on wood substrates			D2793-69(1987)
Print free test	Part C8:1978(1985)	≡ 3678:1976	
Print resistance of lacquers			D2091-88
Compatability of a product with surface to be painted	Part C9:1982(1989)	≡ 4627:1981	
Pressure mottling and blocking resistance of organic films on metal			D3003-71(1987)

Optical tests on paint films	BS 3900	ISO	ASTM
Selecting of coating specimens for appearance measurements			D3964-80(1985)
Colour – visual comparison of	Part D1:1978(1986)	≡ 3668:1976	D1729-82
– and colour difference (determination of principles)	Part D8:1986	≡ 7724/1:1984	
– and colour difference (measurement)	Part D9:1986	≡ 7724/2:1984	
– and colour difference (calculation)	Part D10:1986	≡ 7724/3:1984	D2244-85
– and gloss tolerances of opaque materials, selecting, defining and evaluating conformance			D3134-74
– change in (evaluating with a grey scale)			D2616-88
– permanence of white architectural enamels			D1543-86
– specifying by the Munsell system			D1535-80
– of clear liquids (estimation by the platinum–cobalt scale)		6271-1981	D1209-84(1988)
Metamerism – visual evaluation of			D4086-86
Contrast ratio (hiding power), comparison of, of paints of same type and colour	Part D4:1974(1989)	= 2814:1973	
– of light coloured paints at fixed spreading rate (using polyester film)	Part D6:1982(1985)	≡ 3906:1980	
– of light coloured paints at fixed spreading rate (using black and white charts)		3905:1980	
Hiding power – of white and light coloured paints by the Kubelka–Munk method	Part D7:1983(1989)	≡ 6504/1:1983	
– of paints			D2805-88
– relative, of paints by the visual evaluation of brush outs			D344-84
Gloss – specular of non metallic paint films at 20°, 60° and 85°	Part D5:1980(1987)	≡ 2813:1978	D523-85
– 45 deg 0 deg, directional reflectance factor of opaque specimens			E97-82(1987)
– or sheen uniformity, evaluation of			D3928-80(1987)
– differences between surfaces of similar appearance			D4449-85
– of high gloss surfaces using abridged goniophotometry			E430-78(1983)
Reflection haze of high gloss surfaces			D4039-87
Retro reflectance of horizontal coatings			D4061-81

Mechanical/physical tests on paint films	BS 3900	ISO	ASTM
Abrasion resistance of organic coatings – by air-blast abrasive			D658-81(1986)
– by falling abrasive			D968-81(1986)
– by Taber abrasive			D4060-84
wet, of interior paints			D4213-87
Scrub resistance of interior latex flat wall paints			D2486-79
Washability properties of interior architectural coatings			D3450-86
Washability, practical of organic coatings			D4828-88
Adhesion – by tape test (cross cut test)	Part E6:1974(1989)	= 2409:1972	D3359-87
– of organic coatings by scrape adhesion			D2197-86
– pull-off strength of coatings	Part E10:1979(1989)	≡ 4624:1978	D4541-85
Bendability of pipe-line coatings			G10-83(1988)
Bend test – by Mandrel (cylindrical)	Part E1:1970(1987)	= 1519:1973	D522-88
– by Mandrel (conical)	Part E11:1985	≡ 6860:1984	
Flexibility, coating, of prepainted sheet			D4145-83
Formability – of attached organic coatings with impact-wedge bend apparatus			D3281-84
– of zinc rich primer/chromate complex coatings on steel			D4146-83
Cupping test	Part E4:1976(1989)	=1520:1973	
Indentation – Buchholz test	Part E9:1976(1983)	=2815:1973	
– spherical or pyramidal	Part E12:1986	≡ 6441:1984	
Impact resistance – falling weight method	Part E3:1973/(1985)	≠ TR6272:1979	G14-88
– rapid deformation, resistance to the effects of			D2794-84
– falling ball test	Part E7:1974(1986)		
– pendulum test	Part 8:1974(1979)		
Hardness – scratch test	Part E2:1970(1986)	=1518:1973	
– pendulum damping test	Part E5:1973(1985)	=1522:1973	D4366-87
– indentation of organic coatings			D1474-85
– pencil test			D3363-74(1980)
Chipping resistance of coatings			D3170-87
Static friction, of coating surfaces, measuring			D4518-87
Tensile properties of organic coatings			D2370-82(1987)
Wear resistance of traffic paint, evaluating degree of			D913-88

Chemical resistance tests on paint films	BS 3900	ISO	ASTM
Acid and mortar, resistance of factory applied clear coatings on extruded aluminium			D3260-82(1987)
Chemical agents, effect on organic finishes, used in the transportation industry			D1540-82(1987)
Chemicals, household, effect on clear and pigmented organic finishes			D1308-87
Detergent resistance of organic coatings			D2248-73(1982)
Liquids, resistance to, determination of	Part G5:1976(1984)	=2812:1974	
Stains and reagents, resistance of factory applied coatings on wood			D3023-88
Corrosive environments, evaluation of painted or coated specimens to			D1645-79(1984)
Corrosion resistance, dynamically testing coated steel specimens			D2933-74(1986)
Filiform corrosion resistance of organic coatings on metal	Part F13:1986	≡ 4623:1984	D2803-82(1987)
Humidity resistance – determination of (cyclic condensation)	Part F2:1973(1989)		
– determination of (continuous condensation)	Part F9:1982(1985)	≡6270:1980	
– humid/dry cycling for coatings on wood			D3459-87
– humidity/thermal cycle cracking – on primed metal			D2246-87
– to atmospheres containing sulphur dioxide	Part F8:1976(1986)	=3231:1974	
Water resistance – testing using controlled condensation			D4585-87
– testing using immersion		1521:1973	D870-87
– testing using 100% relative humidity			D2247-87
– testing using water fog apparatus			D1735-87
Salt spray – continuous, resistance to	Part F4:1968(1985)		B117-85
– neutral, resistance to	Part F12:1985	≡ 7253:1984	
Cathodic disbondment – in marine environments (resistance to)	Part F10:1985		
– of coatings for use on land-based buried structures	Part F11:1985		

Durability testing on paint films		BS 3900	ISO	ASTM
Natural weathering	– notes for guidance on conduct of tests	Part F6:1976(1984)	≠ 2810:1974	D1006-73(1986)
	– accelerated outdoor exposure tests			D4141-73(1987)
	– durability of varnishes, exterior			D1641-59(1987)
	– durability and compatibility of factory primed wood products			D2830-82(1987)
	– conducting exterior exposure tests of paints on steel			D1014-83(1988)
Artificial weathering	– enclosed carbon arc type (resistance to)	Part F3:1971(1986)		D822-86
	– fluorescent UV (condensation type)			G53-88
	– unfiltered open flame carbon arc type using dew cycle			D3361-87
	– xenon arc type, with and without water			G26-88
	– fluorescent UV (condensation light and water)			D4587-86
Exposure tests	– single and multi-panel forms for recording results			D1150-55(1987)
	– reporting paint film failures (characteristic of exterior latex paints)			D1848-88
Light fastness	– of paints for interior use to artificial light sources	Part F5:1972	≠ 2809:1976	
Heat	– determination of effect of	Part G7:1976(1985)	= 3248:1975	D2485-84
Dirt collection – on exterior white coated panels				D3719-87
Microbial (fungal or algal) growth – determining presence				D4610-86
Microbial growth or soil and dirt accumulation – evaluation of degree of disfigurement				D3274-82(1988)
Fungal growth	– assessment of resistance to	Part G6:1989		D3456:86
	– on surface of interior coatings in environmental chamber (resistance to)			D3273-86

Miscellaneous test methods	BS 3900	ISO	ASTM
Wood panels for weathering tests of coatings			D358-83(1988)
Panel preparation for paint testing – general	Part A3:1986	≡ 1514:1984	
– wood			D358-83(1988)
– concrete and masonry			D1734-63(1980)
– galvanised steel			D2201-65(1987)
– steel			D609-73(1980)
– aluminium and aluminium alloy			D1730-67(1984)
– aluminium surfaces – hot dip			D1731-67(1984)
– glass			D3891-80
– magnesium alloy			D1732-67(1984)
– zinc coated (galvanised) steel			D2092-86
Determination of quantity of material in a container	Part A15:1976(1985)	= 3232:1974	
Defects – common types of, designation of intensity, quantity and size	Part H1:1983(1989)	≡ 4628/1:1982	
Photographic standards – bleeding of traffic paints (evaluating degree of)			D868-85
– blistering of paints (evaluating degree of)	Part H2:1983(1989)	≡ 4628/2:1982	D714-87
– chalking of exterior paints (evaluating degree of)			D659-86
– checking of exterior paints (evaluating degree of)			D660-87
– cracking of exterior paints (evaluating degree of)	Part H4:1983(1989)	≡ 4628/4:1982	D661-86
– erosion of exterior paints (evaluating degree of)			D662-86
– flaking/(scaling) of exterior paint (evaluating degree of)	Part H5:1983(1989)	≡ 4628/5:1982	D772-86
– rusting of painted surfaces (evaluating degree of)	Part H3:1983(1989)	≡ 4628/3:1982	D610-85
– pictorial surface preparation standards for steel			D2200-85
Efflorescence of interior wall paints			D1736-79(1984)
Acid and base milliequivalent content of electrocoat baths			D4370-84
pH, measuring apparent, of electrocoat baths			D4584-86
Porosity of paint films			D3258-80(1987)
Water vapour permeability of organic coating films			D1653-85
Fire retardency of paints (cabinet method)			D1360-79(1984)
Fire-retardant paints, small scale evaluation (2 foot tunnel method)			D3806-79(1984)

Appendix IV Health and safety legislation

Compiled from information provided by Ken Arbuckle

Introduction

With ever increasing concern for the environment and for the health and safety of individuals, it is inevitable that we shall find more controls and legislation affecting the way in which we formulate and label our products. Each country is developing its own means of controlling risk, through regulations and guidelines, and these often vary in detail from one country to another. It is important for formulators to ensure that they are aware of all legislation, controls and recommendations affecting their products within their specific markets, because to a large extent these factors govern the development of new products.

This appendix outlines some of the more important aspects of health and safety as applied to the paint and allied industries. Sources of information, i.e. publications, major legislation, including contact addresses from where further information can be obtained, and associations and regulatory bodies are listed. The legislation is continually being updated, and it is essential to read the original documents together with any updates in order to obtain a full understanding of the requirements of any act or regulation.

Sources of information

USA

International Civil Aviation Organisation (ICAO) Technical Instructions
International Maritime Dangerous Goods (IMDG) Code
US Federal Register (29CFR 1919, 1200, 48 Federal Register 538280)

Journals

Journal of Protective Coatings and Linings, Technology Publishing Co., 24 S. 18th Street, Pittsburgh, PA 15203. Tel. (412)578 3458

American Paint and Coatings Journal, American Paint Journal Co., 2911 Washington Avenue, St. Louis, MO 63103. Tel. (314) 530 0301

Chemical Abstracts (*CHEMABS*) American Chemical Society, Chemical Abstracts Service, Box 3012, Columbus, OH 43210. Tel. (614) 421 3600.

Journal of Coatings Technology, Federation of Societies for Coatings Technology, 1315 S. Walnut Street, Ste. 830, Philadelphia, PA 19107. Tel. (215) 545 1506

Modern Paint and Coatings, Communication Channels Inc., 6255 Barfield Road, Atlanta, GA 30328. Tel. (404) 256 9800

Paint and Coatings Industry, Western Trade Publishing Co., 7335 Topanga Canyon Boulevard, Ste. 209, Canoga Park, CA 91303

TOXLINE (US National Library of Medicine)

UK/EEC

An Introduction to Local Exhaust Ventilation (HSE Booklet HS(G)27 revised, HMSO)

Approved Codes of Practice and Regulations (HMSO)

Classification, Marking and Packaging of Dangerous Goods (PIRA)

European agreement concerning the international carriage of dangerous goods by road (ADR) (HMSO)

Guidance for the testing of chemicals for toxicity (HMSO)

Hazards, Pollution and Legislation in the Coatings Field, Pergamon, Oxford

Inventory of Existing Chemical Substances (EINECS) Commission of the European Community 1990 (obtainable from (HMSO)

Labelling Guidelines, Paintmakers Association

Orange Book (HMSO)

Registry of Toxic Effects of Chemical Substances, WSCA, (PRA, London)

Regulations concerning the international carriage of dangerous goods by rail (RID) (HMSO)

Journals

Chemistry and Industry, Society of Chemical Industry, 14 Belgrave Square, London SWIX 8PS

Farbe und Lack, Curt R. Vincentz Verlag, Schiffgraben 41-43, Postfach 6247, 3000 Hannover, Germany

Oil and Colour Chemists Association Journal, Oil and Colour Chemists' Association, Priory House, 967 Harrow Road, Wembley HAO 25F

Polymers, Paint and Colour Journal, Fuel and Metallurgical Journals Ltd, Queensway House, 2 Queensway, Redhill, Surrey RHI IQS

Revista del Colore, Rivista del Colore srl, Via degli Imbriani 10, 20158 Milan, Italy

World Surface Coatings Abstracts, Paint Research Association, Pergamon Press Ltd, Headington Hill Hall, Oxford OX3 OBW

Other countries

Chemical Weekly, Sevak Publications, 306 Shri Hanuman Industrial Estate, GD Ambekar Road, PO Box 7110, Wadala, Bombay 400 031, India.

Paintinda, Colour Publications Pvt. Ltd, 126-A Dhurunadi, Off. Drive Nariman Road, Bombay 400 25, India

Surface Coatings Australia, Oil and Colour Chemists Association of Australia, 13 Melby Avenue, Balaclava, Vic. 3183, Australia

Legislation

USA

Major US federal legislation is listed in this section. State regulations are based on this legislation and are reviewed under the following main headings:

Air quality	Transportation
Hazardous wastes	Waste quality
Underground storage tanks	Right to know
Occupational safety and health	Labelling
Toxic substances	Water quality
Packaging	Graffiti

The extent of controls varies from state to state and is constantly changing, making general review impractical.

Classification of hazardous materials

Toxic Substance Control Act (TSCA)

Contact: TSCA Assistance Office (TS799), Office of Toxic Substances, US Environmental Protection Agency, Room EB44, 401 M. Street, S.W., Washington D.C. 20460. Tel. (202) 554 1404

Pollution control, including waste reduction and disposal

Clean Water Act (CWA)

Contact: Office of Drinking Water, US EPA, 401 M. Street, S.W., Washington D.C. 20460

Clean Air Act (CAA)

Contact: Office of Air Quality Planning and Standards, Air Quality Management Division, Research Triangle Park, NC 27711. Tel. (919) 541 5505

Resource Conservation and Recovery Act (RCRA)

Contacts: Office of Solid Waste, US EPA, 401 M. Street. Tel. (202)382 4740; Office of Underground Storage Tanks, U.S. E.P.A., 401 M. Street; RCRA/ Superfund Hotline, Tel. (800) 424 9346 or (202) 382 3000

Superfund/Comprehensive Environment Response, Compensation and Liability Act (CERCLA)

Superfund Amendments and Reauthorisation Act (SARA)

Contacts: Office of Waste Programs Enforcement, US EPA, 401 M. Street; Response Standards and Criterion Branch, US EPA, 401 M. Street

Emergency Planning and Community Right to Know Act (EPCRA)

Contacts: Chemical Emergency Preparedness and Prevention Office (Office of Solid Waste and Emergency Response), US EPA, 401 M. Street; Emergency Planning and Community Right to Know Information Hotline, Tel. (800) 535 0202

Waste Reduction Act (HR1457)

Indoor Air Quality Act (HR1530)

Environmental Crimes Act (HR3641)

Transportation

Hazardous Materials Transportation Act (HMTA)

Contact: Office of Hazardous Materials Transportation, Department of Transportation, 400 Seventh Street, S.W., Washington D.C.20590. Tel. (202) 366 4723

Sanitary Transportation of Food Act (STFA)

Occupational health hazards

Occupational Safety and Health Act (OSHA)

Contact: Office of Public Health/Public Affairs/Information and Consumer Affairs, Occupational Safety and Health Administration, Department of Labor, 200 Constitution Avenue, N.W., Washington D.C. 20210. Tel. (202) 523 8151/9058

Other regulations

Consumer Product Safety Act

Los Angeles Rule 66

National Emission Standards for Hazardous Air Pollutants

OSHA Hazard Communication Rule (1983)

United Nations Recommendations

UK/EEC

Air Quality Standards Regulations (1989) No.317

Control of Air Pollution (Registration of Works) Regulations (1989) No.318

Control of Lead at Work Act (1980) No.1248

Control of Pesticides Regulations (1986) No.1510

Control of Pollution Act (1974) Chapter 40

Control of Pollution (Special Waste) Regulations No.1709

COSHH Regulations (1988) No.1657

CPL (Amendment) Regulations (1990)

Dangerous Substances (Conveyance by Road in Road Tankers and Tank Containers) (1981) No.1059

Health and Safety at Work Act (1974)

Highly Flammable Liquids and Liquified Petroleum Gases Regulations (1972) No.917
Merchant Shipping (Dangerous Goods) Regulations (1981) No.1747
Notification of New Substances Regulations (1982) No.1496
Petroleum (Consolidation) Act (1928)
Road Traffic (Carriage of Dangerous Substances in Packages etc.) Regulations (1986) No.1951 and Amendment (1989) No.105

Associations and regulatory bodies

USA

Air Resources Board (California)
Centre for Food Safety and Applied Nutrition, Food and Drug Administration (FDA), 200 C Street, SW, Washington D.C. 20204. Tel. (202) 472 5690
Consumer Product Safety Commission
Environmental Protection Agency
Federation of Societies for Coatings Technology, 1315 Walnut Street, Philadelphia PA 19107, USA
National Paint and Coatings Association
National Toxicology Programme, Public Information Office MD B2- 04, PO Box 12233, Research Triangle Park, NC 27709. Tel. (919) 541 3991
Occupational Safety and Health Administration
Society for Surface and Protective Coatings

UK/EEC

CERIPAC, 181-203 Avenue Jean Jares, BP 7035, 69007 Lyon, France
CORI, Avenue Pierre Holoffe, B-1342 Limelette, Belgium
EOLAS, Glasnevin, Dublin 9
FPL, Allmandring 37, D-7000 Stuttgart 80, Germany
Health and Safety Executive (HSELINE), UK
Oil and Colour Chemists Association, Priory House, 967 Harrow Road, Wembley HA0 25F
Paintmakers Association, Alembic House, 93 Albert Embankment, London SE1 7TY
Paint Research Association, Waldegrave Road, Teddington, Middlesex TW11 8LD
TNO, PO Box 71, 2600 AB Delft, The Netherlands

Glossary

Compiled by Len Tysall (adapted with permission from Information Brief No. 8, Supplement to Progress Report No. 15, Autumn 1988, Paint Research Association)

To facilitate cross-referencing all entries in this glossary are given in *italic*. Reference should also be made to ASTM-D 16-84, ISO 4617-1-4, List of equivalent terms, ISO 4618-1-3: 1984 Vocabulary and BS 2015: 1965 (1985) Glossary of Paint Terms.

Acid catalysed finishes (acid cat) A type of two-pack paint which cures at ambient temperature and which depends on the catalytic effect of acids. Usually based on a mixture of alkyd and amino resins – much as in a *stoving synthetic* (small additions of acid catalyst can be made to most stoving finishes to lower stoving temperatures – acid cat is an extension of this practice).

Acid value (or acid number) The concentration of free acid groups in a substance, usually in a resin. Expressed in mg KOH per gram.

Alkyd resin A *polyester resin* modified with a fatty acid. Thus, to take a simple example, a polyester resin might be made by reacting glycerol with phthalic anhydride, while an alkyd resin would be made by reacting glycerol with phthalic anhydride and linseed fatty acids. See *Oil free polyester, polyester resin.*

Amine adduct A resinous amine designed to be used as a curing agent for epoxy resins. An amine adduct is made by reacting excess amine with an epoxy resin – the proportions do not therefore lead to complete cure but to a limited reaction. In practice there are two types of amine adduct. Solid adducts are made in a chemical plant using a large excess of amine, the surplus being removed by vacuum distillation. *In-situ* adducts are made by mixing appropriate solutions of resin and amine, the proportions being a compromise between too little amine which would lead to too much reaction and too much amine which would leave to too much unreacted amine. The attraction of *in-situ* adducts is the lower manufacturing cost.

Amino resin A somewhat inaccurate general term for resins made by reacting formaldehyde with an amine or an amide. The two important examples are urea-formaldehyde and melamine-formaldehyde resins.

Anatase Titanium dioxide occurs in three crystal forms; anatase is one of the two which are used as pigments. The other is *rutile.*

Atomisation The breaking up of liquid paint into fine droplets during spray painting.

Bénard cells A name sometimes used to describe the convection cells set up during the formation of paint films by the evaporation of solvents.

Bleeding Staining of a paint film by colour from a previous coat of paint. A typical situation in which this could occur would be when overcoating red paint with white paint (many red pigments are sensitive to solvents). The term is not confined to red staining – it would for instance include the brown stains obtained when overcoating a film of bitumen paint.

Blocked isocyanate The reaction product of an isocyanate and a blocking agent which is usually an alcohol or a phenol or a secondary amine. The reaction is reversed – typically at about 160°C. Blocked isocyanates are used in stoving systems – the free isocyanate released by deblocking reacts with hydroxy groups on a resin. Stoving systems of this type are prone to *gas checking* because cure can take place only at the surface of the film where the isocyanate is released. See *polyurethane*.

Binder A general term for the polymer system of a paint; it covers both resins and oils – also plasticisers. Named 'binder' because it binds the pigment particles together. See *medium, vehicle*.

Bitty Describes the appearance of a film whose surface shows coarse particles of pigment or of gelled resin.

Black Japan The traditional name for a black stoving *enamel*. It was originally an imitation of decorative coatings produced in the Far East (this was also called Oriental lacquer).

Bloom A separate layer on the surface of a film which reduces the gloss. May be caused by syneresis – incompatible material forced out of a resin gel as cure proceeds; an example of this is the amine bloom produced by poorly formulated two-pack epoxies. But bloom on varnish films has been shown to be caused by ammonium sulphate formed at high relative humidity from atmospheric ammonia and sulphur dioxide.

Blushing Opalescence developing in a paint film during drying. Most commonly due to condensed atmospheric water in the film which precipitates the resin from its solution in solvent (water blush); this is most likely to happen during spray painting. Blushing can also be due to badly formulated solvent mixtures which fail to keep the resin or resin mixture in solution; in nitrocellulose lacquers this may involve precipitation of the resin while the nitrocellulose itself is unaffected – this is therefore called resin blush. But in the reverse situation nitrocellulose is precipitated – this is called cotton blush because nitrocotton is an alternative name for nitrocellulose. In a clear film blushing gives a milky appearance but pigmented films tend to turn white irrespective of their normal appearance.

Chalking A thin powdery layer formed on the surface of a film after exposure to weathering; due to breakdown of the resin by UV light. Pigments have a big influence on chalking tendencies – early grades of rutile TiO_2 were particularly bad but chalk-resisting grades have now been developed. A limited degree of chalking may be an advantage in some circumstances because it prevents dirt pick-up – giving self-cleaning paints. Black paint films show a similar breakdown defect; see *sooting*.

Cheesy Soft; applied to a film which has failed to harden properly.

Cissing A film fault in which small circular areas of the substrate are left almost completely bare. This arises from incomplete wetting and may be due to a bad formulation or to ineffective degreasing. *Crawling* is a related more serious fault – here larger irregularly-shaped areas are almost completely bare.

Clear An unpigmented coating – for instance in the currently fashionable car finish 'clear over base'. Note that in the paint field this word has become a noun. See *lacquer, varnish*.

Cold checking A term which is usually applied to wood finishes and refers to the formation of cracks in a film on wood which has been subjected to cold cycling (alternation of hot and cold conditions). It means that the film is not sufficiently flexible. The wood expands much more across the grain than along the grain so that the pattern of cracks tends to follow the grain pattern. It is important to realise that some woods expand much more than others so that consistent results can only be obtained in laboratory tests if the wood panels are carefully selected.

Conventional coating A coating whose films are formed by the evaporation of solvents. Most traditional paint systems are of this type – but most recently developed ones form their films by other mechanisms and are therefore described as *non-conventional*. Conventional coating is a clumsy term but it has a clearly defined meaning. *Solvent-based coating* is an alternative term, but on balance a less satisfactory one.

Convertible coating A coating whose resin system has a curing mechanism. The properties of the film therefore change soon after the film is formed. For example, the film will usually become insoluble in the solvents used in making the paint and recoating problems due to interaction with a previous coat are thus avoided. See *non-convertible coating*.

Covering power A term which is best avoided because it has two possible meanings:

(a) The area which can be coated by a standard volume of paint — this is better called the *spreading power* and would usually be measured in m^2/l.
(b) The ability to hide or obscure the background – this is better called the *hiding power* or better still the *opacity*.

Crawling A film fault in which large irregularly shaped areas of a substrate are left almost completely bare. See *cissing*.

Critical pigment volume concentration (CPVC) The *pigment volume concentration (PVC)* at which most film properties show either maximum or minimum values, or a maximum rate of change. For instance, water permeability shows its lowest (and therefore its most desirable) value at the CPVC. Gloss shows its most rapid rate of change at this point – this is where such terms as *satin, eggshell*, etc are used. The CPVC corresponds to the situation in which resin and pigment are in balance in the sense that there is *just* enough resin to wet the pigment particles and to fill the interstices between them. The quantities of resin and pigment needed will depend not only on the resin and pigment chosen but also on the extent to which the pigment is dispersed (the *degree of dispersion*).

Enamel To a purist there is only one material which should be called an enamel – 'vitreous enamel' which is a specially formulated glass used as a coating. Vitreous enamel is used on the high temperature parts of domestic cookers. It has very high heat, chemical and solvent resistance, is very hard and has excellent colour; but it is rather brittle and expensive in fuel (it is fused at about 850°C). For many applications where very high heat resistance is not needed vitreous has been replaced by stoving paints which are therefore often called 'stoving enamels'. The word enamel is now very widely used in the paint field – so widely that it lacks a precise meaning. Rather vaguely it means a high quality finish and especially one which has a low PVC and therefore gives high gloss and probably requires an undercoat. See *Black Japan*.

Epoxy ester The reaction product of an epoxy resin and a fatty acid. An epoxy ester is a type of oil paint with a general resemblance to an *alkyd resin*. See *oil length*.

Extender A white powder made from a material which has a relatively low refractive index – not very different from that of a typical resin. Calcium carbonate is an example. Such a material will not give any useful degree of opacity but it will produce the same effects as a pigment in other ways – for example, it will reduce gloss or increase the hardness of a film. It can therefore be a valuable raw material, and is usually much cheaper than a pigment.

Fineness of grind Another name for *degree of dispersion*, but a less desirable one because 'grind' suggests that the pigment particles are being broken, whereas the problem is to separate them.

Flat Describes a film with low gloss – as for instance that obtained from a 'flat oil paint'. Flat is best avoided because it invites confusion with the operation of *flatting* and an alternative word is *matt*.

Flatting (or sanding) Abrasion of a paint film intended to produce a level surface prior to applying finishing coats.

Floating/flooding Partial separation of two different pigments during film drying. This can result either:

(a) in a uniform change of colour during drying so that the film colour is different from that of the wet paint. Thus for example grey films nearly always 'dry dark'.

or

(b) in a non-uniform change giving a convection cell pattern (which will be distorted if the film was brush applied).

There seems to be some disagreement as to which of these effects if *floating* and which is *flooding*.

Flocculation Spontaneous reversal of pigment dispersion – must be prevented to produce a usable paint. This requires the presence of a suitable surface active agent. Many resins can perform this function but some may require an addition of a dispersing aid.

Flow control agent An additive which decreases the likelihood of *cissing*. The term seems to be applied only to two-pack systems.

Gas checking Wrinkling of a stoved film giving a variety of patterns according to conditions – some patterns are so fine that they can only be recognised under magnification; others follow the convection cell pattern. This wrinkling is caused by the upper part of the film curing more rapidly than the lower part. Solvent from below then attacks the gel formed by partial cure, the gel swells and therefore piles up in wrinkles. It is more liable to happen in a gasfired oven (but it can happen in other ovens). Gas checking in stoved oil paints is due to oxygen being available only at the film surface. Most modern stoving paints cure by a different chemical mechanism – polyetherification – and would not therefore respond to oxygen in this way. They might respond to acidic impurities in the atmosphere; however, this is generally prevented by acid groups on one of the resins which ensure a fairly uniform cure from top to bottom of the film. Stoving systems based on *blocked isocyanates* are very likely to show gas checking. Systems which show gas checking can be used in heavily pigmented paints such as primers or undercoats, or in dark finishes where thin films will give acceptable opacity. See *Black Japan*.

Gloss The reflectivity of the surface of a film. Gloss meters measure this with photocells using a black glass panel as standard (angles of incidence and reflection should be quoted). *Gloss*, like colour, is not just a physical measurement but involves psychological factors. A number of terms are used to describe gloss in the critical semi-gloss region. These include *sheen, satin, silk, eggshell, semi-gloss* and *semi-matt* (this is an approximately decreasing order of gloss but there are no formal definitions).

Glossing up Increase in gloss of a matt film in service due to handling and other casual contacts. Matt paints must be formulated so as to minimise this effect.

Grinning through Details of the background which can be seen through the film are said to be 'grinning through'.

Guide coat A thin coat of paint which is completely rubbed away during a flatting operation. The point being it ensures all the original paint film has been abraded.

Hiding power See *opacity*.

Holidays Areas which have been missed during a painting operation.

Hungry Used to describe a surface which is very porous and said to have high *suction*. Such surfaces will soak up large amounts of paint without showing an acceptable 'painted' appearance. See *suction*.

Hydroxyl value (or hydroxyl number) The concentration of free hydroxyl groups on a substance, usually on a resin. Expressed in mg KOH per gram.

Lacquer A troublesome word because it has two meanings:

(a) A clear coating, i.e. one containing no pigment.
(b) A coating that dries without chemical cure.

To avoid possible confusion it is best to use *clear* for (a), as in the car finish 'clear over base'; and *non-convertible* for (b). However, *non-convertible* is a clumsy word and it is almost impossible to avoid *lacquer* when referring to nitrocellulose-based paints because the term *nitrocellulose lacquer* is so firmly established. (Note that meaning (a) is shared with the term *varnish*.)

Latex paint An alternative name for an *emulsion paint.*

Laying off The final operation in brush painting in which the whole area is lightly stroked with the brush in one direction.

Leafing The process in which plate-like pigments such as aluminium or micaceous iron oxide arrange themselves more or less parallel to the film surface. Effective leafing improves film appearance in metallics and also improves water resistance. Effective leafing is favoured by heavy *shrinkage.*

Lifting Swelling of an apparently dry coat of paint when a second coat (possibly of the same paint or possibly of a different one) is applied. This is due to solvent action on an incompletely cured resin – physical tests of drying can be misleading as a guide to degree of cure. The practical result of lifting is more or less complete loss of adhesion. In serious cases therefore it may be necessary to strip off all paint and start again.

Mar resistance The ability to resist minor surface scratching in service. Mar resistance is not directly related to film hardness; it can sometimes be improved by small additions of wax.

Matt Describes a film with very low gloss. See *flat, vinyl matt.*

Medium A general term for the continuous phase in which the pigment is dispersed. This phase will be called the *vehicle* in a liquid paint but the *binder* in a dry film.

Moisture set The usual name for a type of *polyurethane* which relies on reaction with atmospheric moisture for its curing mechanism.

Nibs The small particles of dust which are visible on the surface of a dry film; also called 'bits'. The removal of these nibs with abrasive paper is called *denibbing.*

Nitrocellulose lacquer A *lacque*r in which nitrocellulose is the *binder* or part of the binder. Note that nitrocellulose is a chemical misnomer; the material is a (partial) nitrate of cellulose and *not* a nitro-compound. But the name nitrocellulose is so firmly established – even in official regulations – that there is no prospect of eliminating it.

Non-aqueous dispersion (NAD) A dispersion of a resin in an aliphatic solvent. Made by a process analogous to emulsion polymerisation – special surface active agents are necessary. Most commonly used with acrylic resins which give very low solids in solution. The technique makes it possible to produce high solids/high molecular weight paints based on acrylic resins.

Non-conventional coating A term sometimes used to define a coating *not* dependent on solvent evaporation for its film formation (see *conventional coating*). Most of the paint systems developed in recent years are non-conventional. Probably the best known example is the *emulsion paint* which forms its films by the coalescence of resin particles. Note that a *non-conventional coating* may contain solvents but these solvents will not play a major part in film formation.

Non-convertible coating A coating whose resin system has no curing mechanism. There is thus no change in properties during film drying. Problems are commonly encountered when applying later coats of paint because of attack by solvent. See *convertible coating, lacquer.*

Non-volatile material (NVM) The proportion of a total paint formulation which does not evaporate during the film-forming process – normally expressed as a percentage. *NVM* is popularly called *solids* but this is a misleading term and must not be taken literally. See *plasticiser.*

Oil-free polyester A polyester which does not include a fatty acid as part of its raw materials. If the definitions of *polyester* and *alkyd resin* are accepted there is no point in saying 'oil free'. To add to the confusion it is not unusual to hear the name *oil-free polyester* applied to a combination of polyester and *melamine resin. Stoving polyester* is often used as an alternative to *oil-free polyester.* Oil-free polyesters compete with *stoving synthetics* and with *thermosetting acrylics.*

Oil length A measure of the balance between oil and resin components in an oleoresinous varnish, an alkyd, a urethane oil or an epoxy ester. The principle is the same in all these cases but the oil length is expressed in different ways. Thus:

(a) For an *oleoresinous varnish* the oil length is the weight ratio between oil and resin. A long oil varnish will have an oil length greater than 2½:1; a short oil varnish less than 1¼:1. The American system uses the number of (US) gallons of oil per 100 pounds of resin – long oil is greater than 25 gallons (per 100 pounds resin), short oil less than 10 gallons. The two sets of definitions are roughly in line.
(b) For an *alkyd resin* one method is to define oil length as the percent oil because there is about 10% loss of water during processing it is important to state the basis of the calculation – this is normally the *theoretical yield.* A second method uses percent fatty acids instead of percent oil, but is otherwise the same. Conversion can be made from one basis to the other – the two measures are not greatly different – for example, 45% oil is approximately equal to 40% fatty acids. Long oil is usually taken as $> 60\%$ and short oil as $< 40\%$ (whether on oil or fatty acid basis).
(c) Oil length of a *urethane oil* is usually expressed in the same way as for alkyds.
(d) There is no formal definition of oil length of an *epoxy ester.* The usual practice is to refer to the percentage of (hydroxy plus epoxy) groups which have been esterified, commonly expressed by a code number – for example, '4' means 40% esterified and so on.

Oleoresinous varnish A combination of oil and resin normally 'cooked' together at a high temperature. This is the traditional varnish – the more cumbersome name is used to clarify the double meaning of *varnish.*

Opacity The ability to obscure background. Opacity is usually tested by applying controlled amounts of paint over standard black and white cards. *Opacity* is preferred to *covering power.*

Organosol A paint based on PVC or a related resin dispersed, together with a pigment, in a plasticiser and a substantial amount of solvent. Like a *plastisol*, an *organosol* forms its film by fusion. Organosols are easier to apply because the lack of restriction on solvent content allows a free choice of viscosity but the high solvent content limits film thickness.

Overspray The paint which misses the target in a spray painting operation.

Pigment volume concentration (PVC) The percentage of pigment in a dry film calculated on a volume basis. Pigment in this sense includes *extender*. See *critical pigment volume concentration, degree of dispersion.*

Plasticiser A liquid which can dissolve a resin but which is non-volatile and becomes part of the dry film. It is therefore included in the *solids* of the paint formulation – solids is more correctly called *non-volatile material*. The difference between a plasticiser and a solvent is a matter of volatility.

Plastisol A paint based on PVC or a related resin dispersed, together with a pigment, in a plasticiser. Small amounts of solvent are sometimes used to reduce viscosity. Plastisols form their films by fusion, typically at about 160°C. See *organosol.*

Polyester resin In purely chemical terms any resin made by esterification, but in the paint field the term is usually confined to resins which do *not* include fatty acids as part of their raw materials. See *alkyd resin, oil-free polyester.*

Polyisocyanate A resinous isocyanate designed to be used as the curing agent in a two-pack polyurethane system. A number of techniques are used – for example, reacting an excess of isocyanate with a triol such as trimethylolpropane or alternatively forming a cyclic trimer of an isocyanate. Polyisocyanates are analogous to the *amine adducts* used in two-pack epoxy systems. See *polyurethane*

Polyurethane A resin system involving urethane groups, either as part of the existing resin structure or formed during a curing process. A number of different urethane paint systems are in use, the most important being:

(a) Two-pack consisting of a resin carrying hydroxyl groups which is cured with a *polyisocyanate*
(b) One-pack *moisture-set* systems in which a specially formulated *polyisocyanate* cures by reaction with atmospheric water
(c) One-pack stoving systems consisting of a resin carrying hydroxyl groups cured at elevated temperatures by a *blocked isocyanate*
(d) *Urethane alkyds* produced by reacting a limited amount of isocyanate with an *alkyd resin*
(e) *Urethane oils* produced by reacting partial glycerides or similar compounds with a major amount of isocyanate.

Powder coating A type of paint in which all the components are solid. Solid in this context includes resins which are more correctly regarded as super-cooled liquids – from a practical point of view the components must be solid enough to give a dry powder which will not block in normal storage. In the paint field only thermosetting powders are of interest – thermosetting powders compete with stoving paints. Thermoplastic powders are made by the plastics industry. They are normally applied in very thick films and do not compete with paints.

Primer The first coat of a painting system. On a non-porous substrate such as a metal the essential function of a primer is to give good adhesion but it will also be expected to help in resisting corrosion on metals such as steel. On a porous substrate adhesion is unlikely to be a problem and the primer's main function will be to act as a sealer – that is, to prevent excessive absorption of later coats of paint.

Rutile Titanium dioxide occurs in three crystal forms; rutile is one of the two which are used as pigments. The other is *anatase*.

Sanding (or flatting) Abrasion of a paint film intended to produce a level surface prior to applying finishing coat(s).

Saponification value (sap value) A measure of the concentration of saponifiable material in a substance, usually in a resin. Expressed in mg KOH per gram; this includes the KOH required to neutralise the free acid groups.

Satin See *gloss*.

Sealer A coating, which may be clear or pigmented, formulated to act as a barrier coat. A sealer might:

(a) prevent *bleeding* or
(b) protect a paint film, or a substrate, from being softened or swollen by solvents in subsequent coats (see *lifting*) or
(c) reduce the *suction* of a surface.

Semi-gloss See *gloss*.

Semi-matt See *gloss*.

Sheariness Uneven gloss on the surface of a paint film.

Sheen See *gloss*.

Shrinkage Loss of solvent from a film leads first to an increase in viscosity and then to the formation of a gel. This gel contains solvent which can escape only by diffusion; loss of solvent causes the gel to shrink. There are important effects arising from this shrinkage; for example, heavy shrinkage gives an attractive appearance to a metallic finish though it has the disadvantage of reducing adhesion to metal surfaces. The extent of shrinkage in a paint system varies widely with the chosen resin and solvent. See *leafing*.

Silk See *gloss, vinyl silk*.

Sleepy Slightly reduced in gloss (applied to a film which was intended to be glossy but which is somewhat below this standard – the term would not be used for a film which had lost almost all its gloss).

Solids Another name for *non-volatile material*, but a misleading one because many of the components of the dry film are not truly solid – for example, *plasticisers* which are clearly liquids, and resins which are correctly regarded as super-cooled liquids.

Solvent-based coating An alternative to the term *conventional coating*. Use of 'solvent-based' is not recommended because there are systems which contain solvent but which do not form their films by the evaporation of that solvent and it is not clear whether these are 'solvent-based' or not. Another difficulty is that there is no clear opposing term with which to contrast 'solvent-based'.

Sooting A thin powdery layer formed on the surface of a black film after exposure to weathering and due to breakdown of the resin by UV light. Sooting is very common with films of bitumen. This behaviour is analogous to the 'chalking' of white films. See *chalking*.

Spreading power (or spreading rate) The area which can be effectively covered by a standard volume of paint, usually expressed in m^2/l. *Spreading power* is preferred to *covering power*.

Starved If part of the area of a painted article has not received its full quota of paint it is said to be 'starved'.

Stopper A paste-like material intended to fill holes and cracks in a surface to be painted. Usually applied by knifing and sanded smooth after drying. See *surfacer*.

Stoving polyester An alternative name for an *oil-free polyester*.

Stoving synthetic A common name for a type of stoving paint in which the binder is a combination of an *alkyd resin* and an *amino resin*. The name could reasonably be applied to all the currently used stoving enamels. Competes with *oil-free polyester* and with *thermosetting acrylic*.

Suction The tendency of a porous surface to absorb paints during application; this makes it difficult to build up an acceptable finish. See *hungry*.

Surfacer A heavily pigmented paint formulated to fill minor surface irregularities – perhaps just the 'grain' of a metal surface. It is applied like a paint, usually by spraying but perhaps by brushing. Usually sanded before finishing coats are applied but there are now some 'non-sand' sealers which do not require this. See *stopper*.

Theoretical yield The weight of resin which would be obtained if the chemical reactions involved were to go to completion. This cannot be achieved in practice but it gives a convenient reference point for calculations. See *oil length*.

Thermoplastic acrylic (TPA) A *non-convertible coating* in which an acrylic resin forms the binder or is a major component of the binder. See *thermosetting acrylic*.

Thermosetting acrylic (TSA) A type of stoving paint in which an acrylic resin is the binder or a major component of the binder. The most widely used type is based on a combination of acrylic resin and *amino resin*. This would compete with *stoving synthetic* and with *oil-free polyester*. See also *thermoplastic acrylic*.

Turpentine substitute (turps sub, mineral turpentine) An early name for white spirit which was first introduced to replace turpentine.

Undercoat In a full painting system an undercoat is applied after the primer and surfacer and before the finish. Its specialised function is to give opacity. It must, of course, have a colour which is close to that of the finish because the finish is formulated mainly to give gloss and cannot therefore be expected to give much opacity.

Urethane alkyd A resin made by reacting an *alkyd resin* with a limited amount of a di-isocyanate – there is always an excess of hydroxyl and carboxyl groups present compared with isocyanate groups so that the free isocyanate content is low. See *urethane oil.*

Urethane oil A resin made by reacting partial glycerides or similar compounds with a major amount of di-isocyanate – this amounts to replacing the phthalic anhydride in a conventional alkyd by di-isocyanate. See *urethane alkyd.*

Varnish A troublesome word because it has two meanings:

(a) A *clear* coating (i.e. one containing no pigment), as in 'a coat of varnish'.
(b) A combination of oil and resin – the traditional binder for 'coach paints' (see *oleoresinous varnish*).

To avoid possible confusion it is best to use *clear* for (a), as in the fashionable car finish 'clear over base'; and *oleoresinous varnish* for (b).
(Note that meaning (a) is shared with the word *lacquer.*)

Vehicle See *medium, binder.*

Vinyl lacquer *A non-convertible coating* in which the *binder* is a *vinyl resin.*

Vinyl matt The popular name for a low gloss *emulsion paint.* 'Vinyl' here refers to vinyl acetate which was the resin used in early emulsion paints in the UK – it is still important in the form of copolymers. See *emulsion paint, vinyl silk, vinyl resin.*

Vinyl resin A copolymer of vinyl chloride and (a minor amount of) vinyl acetate which is the binder in *vinyl lacquers.* There are also some modified grades which include a third monomer – these are called 'terpolymers'. The name 'vinyl' used here is easily confused with *vinyl matt* and *vinyl silk.*

Vinyl silk The popular name for an *emulsion paint* with a moderate level of gloss. See *emulsion paint*, *vinyl matt*, *vinyl resin*, *vinyl lacquer.*

Water-based A general name for resin systems in which solvents have been replaced, or at least largely replaced, by water. Important examples are *emulsion paints, electrodeposition paints* and *water-based* stoving enamels. The name *water-based* is adopted by analogy with *solvent-based.*

Index